Pseudorandomness and Cryptographic Applications

PRINCETON COMPUTER SCIENCE NOTES

David R. Hanson and Robert E. Tarjan, Editors

Pseudorandomness and Cryptographic Applications

Michael Luby

PRINCETON UNIVERSITY PRESS

PRINCETON, NEW JERSEY

ISBN 0-691-02546-0

Library of Congress Cataloging-in-Publication Data

A CIP catalog record for this book is available from the
Library of Congress

The publisher would like to acknowledge the author of this volume for
providing the camera-ready copy from which this book was printed

Princeton University Press books are printed on acid-free paper and meet the
guidelines for permanence and durability of the Committee on
Production Guidelines for Book Longevity
of the Council on Library Resources

Printed in the United States of America
by Princeton Academic Press

1 3 5 7 9 10 8 6 4 2

Table of Contents

Introduction of some basic notation that is used in all subsequent lectures. Review of some computational complexity classes. Description of some useful probability facts.

Introduction to private key cryptosystems, pseudorandom generators, one-way functions. Introduction of some specific conjectured one-way functions.

Discussions of security issues associated with the computing environment of a party, including the security parameter of a protocol. Definition of an adversary, the achievement ratio of an adversary for a protocol, and the security of a protocol. Definitions of one-way functions and one-way permutations, and cryptographic reduction.

Definition of a weak one-way function. Reduction from a weak one-way function to a one-way function. More efficient security preserving reductions from a weak one-way permutation to a one-way permutation.

Proof that the discrete log problem is either a one-way permutation or not even weak one-way permutation via random self-reducibility. Definition of a pseudorandom generator, the next bit test, and the proof that the two definitions are equivalent. Construction of a pseudorandom generator that stretches by a polynomial amount from a pseudorandom generator that stretches by one bit.

Introduction of a two part paradigm for derandomizing probabilistic algorithms. Two problems are used to exemplify this approach: witness sampling and vertex partitioning.

Definition of inner product bit for a function and what it means to be a hidden bit. Description and proof of the Hidden Bit Theorem that shows the inner product bit is hidden for a one-way function.

Definitions of statistical measures of distance between probability distributions and the analogous computational measures. Restatement of the Hidden Bit Theorem in these terms and application of this theorem to construct a pseudorandom generator from a one-way permutation. Description and proof of the Many Hidden Bits Theorem that shows many inner product bit are hidden for a one-way function.

Definitions of various notions of statistical entropy, computational entropy and pseudoentropy generators. Definition of universal hash functions. Description and proof of the Smoothing Entropy Theorem.

Reduction from a one-way one-to-one function to a pseudorandom generator using the Smoothing Entropy Theorem and the Hidden Bit Theorem. Reduction from a one-way regular function to a pseudorandom generator using the Smoothing Entropy Theorem and Many Hidden Bits Theorem.

Definition of a false entropy generator. Construction and proof of a pseudorandom generator from a false entropy generator. Construction and proof of a false entropy generator from any one-way function in the non-uniform sense.

Definition of a stream private key cryptosystem, definitions of several notions of security, including passive attack and chosen plaintext attack, and design of a stream private key cryptosystem that is secure against these attacks based on a pseudorandom generator.

Definitions and motivation for a block cryptosystem and security against chosen plaintext attack. Definition and construction of a pseudorandom

function generator from a pseudorandom generator. Construction of a
block private key cryptosystem secure against chosen plaintext attack
based on a pseudorandom function generator.

Discussion of the Data Encryption Standard. Definition of a pseudorandom invertible permutation generator and discussion of applications
to the construction of a block private key cryptosystem secure against
chosen plaintext attack. Construction of a perfect random permutation
based on a perfect random function.

Construction of a pseudorandom invertible permutation generator from
a pseudorandom function generator. Definition and construction of a
super pseudorandom invertible permutation generator. Applications to
block private key cryptosystems.

Definition of trapdoor one-way functions, specific examples, and construction of cryptosystems without initial communication using a private
line.

Definition and construction of a universal one-way hash function.

Definition and construction of secure one bit and many bit signature
schemes.

Definition of interactive proofs **IP** and the zero knowledge restriction of
this class **ZKIP**. Definition and construction of a hidden bit commitment scheme based on a one-way function. Construction of a **ZKIP** for
all **NP** based on a hidden bit commitment scheme.

Overview and Usage Guide

These lectures stress rigorous definitions and proofs related to cryptography. The basic tenets underlying this development are the following:

(physical assumptions) It is possible to physically protect information that is stored in a single location, but much harder to physically protect information that is sent over long distances.

(randomness assumption) It is possible to generate random and uniformly distributed bits at a single location.

(computational assumptions) There are limits on the amount of computation time that is considered to be reasonable. There are one-way functions, i.e., functions which are easy to compute but hard to invert in a reasonable amount of time.

A basic goal of cryptography is to be able to send information privately between locations that are physically far apart. A protocol achieving this goal can be easily implemented using a pseudorandom generator. The first half of the monograph develops the ideas used to construct a pseudorandom generator from any one-way function. The second half of the monograph shows how to use a one-way function to construct other useful cryptographic primitives and protocols such as stream and block private key cryptosystems, pseudorandom function generators, pseudorandom invertible permutation generators, signature schemes, hidden bit commitment protocols and zero-knowledge interactive proof systems.

———————∞———————

The Preliminaries are meant to introduce basic notation and material that is used throughout the remaining lectures. It is best to initially skim over this, and read it more carefully as the need arises.

Lecture 1 starts with an informal description of the problem of sending private messages on a public line. Solutions to this basic problem (and many other problems in cryptography and other areas as well) can be based on a pseudorandom generator. It turns out that a pseudorandom generator can be constructed from a one-way function, and this construction is the main emphasis of the first few lectures.

Informal notions of a one-way function and a pseudorandom generator are introduced in Lecture 1 and the connection is made between these concepts and the problem of communicating privately using a public line.

Lecture 2 develops more quantitative and definitive formulations of basic concepts related to the security of cryptographic protocols and security preserving properties of reductions.

The first technical results appear in Lecture 3. This lecture shows how to construct a one-way function from a weak form of a one-way function.

Lecture 4 introduces the formal notion of a pseudorandom generator and provides some easy reductions between various forms of pseudorandom generators.

Lecture 5 sets up some of the basic technical ideas that are used in many of the subsequent lectures (and in a wide variety of other problems as well). The ideas developed in this lecture have applications in a number of different areas, and this lecture can be read independently of the material in the other lectures.

Lecture 6 begins in earnest discussing the reduction from a one-way function to a pseudorandom generator. Lecture 7 discusses the relationship between classical notions of statistical distances between distributions and the important notion of computational distance between distributions. Lecture 8 discusses classical notions of entropy and non-classical notions of computational entropy. Lectures 9 and 10 finally culminate with the construction of a pseudorandom generator from any one-way function.

A natural break in the monograph occurs between Lecture 10 and Lecture 11. In Lecture 11 and subsequent lectures, we develop a variety of cryptographic protocols based on primitives introduced and developed in the preceding lectures.

Lecture 11 introduces a notion of a stream private key cryptosystem. This is a system that enables one party to send to another a stream of private bits over a public line after having established a private key. We give a straightforward implementation of such a system based on a pseudorandom generator.

Lecture 12 introduces the notion of a block private key cryptosystem. This type of system is more versatile and easier to use in practice than a stream system. For the purpose of implementing a block cryptosystem, we introduce the notion of a pseudorandom function generator, show how to construct a pseudorandom function generator based on a pseudorandom generator, and then show how to implement a block cryptosystem based on a pseudorandom function generator.

Lectures 13 and 14 introduce stronger notions of block cryptosystems, ones similar to what is used in practice, and show how these can be implemented based on a pseudorandom function generator.

One of the drawbacks of a private key cryptosystem is that it is assumed there is an initialization phase between the two parties where the private key is established in complete privacy. Lecture 15 introduces trapdoor one-way functions and trapdoor one-way predicates, and shows how a cryptosystem can be constructed without this assumption.

Lecture 16 shows how to construct a universal one-way hash function. This is instrumental to the construction of a signature scheme given in Lecture 17. Lecture 17 also shows how to construct a signature scheme. Finally, Lecture 18 briefly touches on the subjects of interactive proofs and zero knowledge, introduces a protocol for bit commitment, and proves that every **NP** language has a zero knowledge proof based on bit commitment.

There are a number of exercises scattered throughout this monograph. In terms of scale of difficulty, some of the exercises can be immediately solved, some exercises do not take a great deal of innovation to solve but do require a fairly good understanding of the basic definitions, while other exercises require both a fairly deep grasp of the definitions and ingenuity. The exercises are not categorized in terms of difficulty in the text. (A loose categerization is given in the List of Exercises and Research Problems.) This is partly because such a categorization is so subjective in the first place, and partly because this gives the exercises more of a research type flavor (in research, the difficulty of the problem is almost never known a priori). There are also a few research problems scattered throughout. Go for it!

Mini-Courses

Here are two suggested mini-courses that could be taught based on parts of the monograph. The first concentrates on how to construct a basic private key cryptosystem from a one-way permutation. The second emphasizes the definitions and uses of cryptographic primitives.

Basic Private Key Cryptography : Preliminaries (as needed), Lectures 1 and 2, Lectures 4-7, Lecture 11.

Definitions and Uses of Cryptographic Primitives : Preliminaries (as needed), Lectures 1 and 2, Lecture 3 (the first two constructions), Lectures 11-18.

Acknowledgments

These lectures developed from a graduate level course taught at U.C. Berkeley in the Fall semester of 1990. I would like to thank the scribes for this course: Steven Procter, HuaSheng Su, Sanjeev Arora, Michael Kharitonov, Daniel Rice, Amit Gupta, David Blackston, Archie Cobbs, Madhu Sudan and Abhijit Sahay. They all provided useful initial drafts of these notes. I would also like to thank Boban Veličković, who was kind enough to give one of the lectures during the course.

Oded Goldreich made many suggestions that tremendously improved these lectures, including pointing me to the simple proof of the Hidden Bit Theorem in Lecture 6, and suggesting simplifications to the proofs and constructions in Lecture 17. Dana Randall caught a number of technical mistakes in the presentation, suggested substantial improvements to my too often convoluted and twisted presentation, and motivated some of the exercises. Bruno Codenotti made a number of helpful suggestions, fixed many typos and suggested some general and specific clarifications to the text. Marek Karpinski made a number of simplifying and clarifying suggestions about notation and definitions. Leonard Schulman taught me some of the inequalities and proofs concerning entropy in Lecture 8. Moni Naor helped to clarify the presentation of the results described in Lectures 16 and 17, and suggested some of the exercises in those lectures. Johannes Blömer helped to write the proofs in Lecture 7 in a style consistent with the proofs given in Lecture 6. Ran Canetti (at 3a.m.) and Rafail Ostrovsky made invaluable suggestions about how to define the attacks against cryptosystems in Lecture 11 and Lecture 12. Moti Yung reviewed the entire monograph, and made numerous suggestions on notation and basic definitions, caught a number of mistakes, suggested some of the exercises and material to be included, and substantially helped to clarify the entire presentation. Shafi Goldwasser made valuable suggestions about the entire manuscript that inspired much rewriting, and she particularly helped with Lecture 15. Rajeev Motwani used a preliminary version of this monograph for parts of a course he taught at Stanford. This provided invaluable feedback and caught numerous glitches in the presentation. I would also like to thank Hugo Krawczyk, Andy Yao, Ron Rivest, Leonid Levin, Charlie Rackoff, Matt Franklin and Johan Håstad for glancing over this monograph and providing me with (in some cases, detailed) feedback.

Proper referencing is as always a sensitive subject. To allow coherent presentation of the material without distractions, I have chosen to place all credits for the material contained in the monograph at the end. It goes without saying that this monograph wouldn't have existed without the prodigious and ingenious work produced by a number of researchers. I have tried reasonably hard to give proper credit to the researchers who have made substantial contributions to the material covered in these lectures. However, I am quite sure there will be some who have made substantial contributions and who did not receive proper credit: my apologies in advance and please let me know for future (if any) versions of this monograph.

I would like to thank the International Computer Science Institute for allowing me to take the time to develop this monograph. In addition, partial support for this work was provided by *NSF* operating grants CCR-9016468 and CCR-9304722 and by *Israeli-U.S. NSF Binational Science Foundation* grants No. 89-00312 and No. 92-00226.

Finally, I would like to thank my parents, and especially my father, for continual inquiries about the status of the monograph and insistent encouragement to stay with it till it was finished, done, out the door.

–Michael Luby

Pseudorandomness and Cryptographic Applications

Preliminaries

Overview

We introduce some basic notation for bits, sets, strings, matrices, functions, numbers and probability. We review some standard computational complexity classes and probability facts.

Basic Notation

In this section, we introduce much of the notation that will be used in subsequent lectures.

$\mathcal{N}$ is the set of natural numbers, and $n \in \mathcal{N}$. $\mathcal{R}$ is the set of real numbers.

Set Notation : We let $\{0,1\}^n$ be the set of all n bit strings, and we let $\{0,1\}^{\leq n}$ be the set of all bit strings of length at most n. If S is a set then $\sharp S$ is the number of elements in S. If T is also a set then $S \setminus T$ is the set of elements in S that are not in T. If p is a positive integer, then

$$\mathcal{Z}_p = \{0,\ldots,p-1\}$$

and

$$\mathcal{Z}_p^* = \{z \in \{1,\ldots,p-1\} : \gcd(z,p) = 1\}.$$

Note that if p is prime then $\mathcal{Z}_p^* = \{1,\ldots,p-1\}$. We can view $\mathcal{Z}_p$ as an additive group and $\mathcal{Z}_p^*$ as a multiplicative group.

String Notation : We let $\|x\|$ denote the length of x. We let $\langle x, y \rangle$ denote the sequence of two strings x followed by y, and when appropriate we also view this as the concatenation of x and y. If $x \in \{0,1\}^n$ then x_i denotes the ith bit of x and $x_{\{i,\ldots,j\}}$ denotes $\langle x_i,\ldots,x_j \rangle$. If $x,y \in \{0,1\}^n$ then $x \oplus y$ is $\langle x_1 \oplus y_1,\ldots,x_n \oplus y_n \rangle$. The string 1^n denotes the concatenation of n ones, and similarly 0^n denotes the concatenation of n zeroes, and λ is the empty string.

Matrix Notation : Let $x \in \{0,1\}^{m \times n}$ be an $m \times n$ bit matrix.

- $x_i \in \{0,1\}^n$ refers to the ith row of x.

- $x_{\{i,\ldots,j\}} \in \{0,1\}^{(j-i+1)\times n}$ refers to rows i through j of x.

- $x_{i,j} \in \{0,1\}$ refers to the (i,j)-entry in x.

We can view x as a string of length mn, which is the concatenation $\langle x_1, \ldots, x_m \rangle$ of the rows of the matrix.

The $\odot$ operation indicates matrix multiplication over GF[2]. If $x \in \{0, 1\}^n$ appears to the left of $\odot$ then it is considered to be a row vector, and if it appears to the right of $\odot$ it is considered to be a column vector. Thus, if $x \in \{0, 1\}^n$ and $y \in \{0, 1\}^n$ then $x \odot y = \sum_{i=1}^n x_i y_i \bmod 2$. More generally, if $x \in \{0, 1\}^{\ell \times m}$ and $y \in \{0, 1\}^{m \times n}$ then $x \odot y$ is the $\ell \times n$ bit matrix, where the (i, j)-entry is $r \odot c$, where r is the ith row of x and c is the jth column of y.

$\{1, -1\}$**-bit notation :** Sometimes, we find it convenient to view bits as being $\{1, -1\}$-valued instead of $\{0, 1\}$-valued. If $b \in \{0, 1\}$ then $\bar{b} \in \{1, -1\}$ is defined to be $\bar{b} = (-1)^b$. If $x \in \{0, 1\}^n$ then $\bar{x} \in \{1, -1\}^n$ is defined as the string where the ith bit is $\bar{x}_i$.

Number Notation : If a is a number then $|a|$ is the absolute value of a, $\lceil a \rceil$ is the smallest integer greater than or equal to a, $\log(a)$ is the logarithm base two of a and $\ln(a)$ is the natural logarithm of a (i.e., the logarithm base $e = 2.71828\ldots$).

Function Notation : Let S and T be sets.

- **Fnc:**$S \to T$ is the set of all functions mapping S to T.

- **Perm:**$S \to S$ is the set of all permutations from S to S.

Probability Notation

In general, we use capital letters to denote random variables and random events. Unless otherwise stated, all random variables are independent of all other random variables. If X is a random variable and f is a function then $f(X)$ is the random variable defined by evaluating f on an input chosen according to X. We use E to denote expected value of a random variable, e.g., $E[f(X)]$ is the expected value of $f(X)$, which is defined in terms of the distribution on X.

We use Pr to denote the probability, e.g., $\Pr[X = x]$ is the probability that random variable X takes on the value x. When S is a set we use the notation $X \in_{\mathcal{U}} S$ to mean that X is a random variable uniformly distributed in S, and $x \in_{\mathcal{U}} S$ indicates that x is a fixed element of S chosen uniformly. More generally, if $\mathcal{D}$ is a probability distribution on a set S, then $X \in_{\mathcal{D}} S$ indicates that X is a random variable distributed in S according to $\mathcal{D}$, and $x \in_{\mathcal{D}} S$ indicates that x is a fixed element of S chosen according to $\mathcal{D}$. If $\mathcal{D}$ is a distribution on a set S and $\mathcal{E}$ is a distribution on a set T then we let $\mathcal{D} \times \mathcal{E}$ be the product distribution of $\mathcal{D}$ and $\mathcal{E}$ on the set $S \times T$, i.e., the random variables $Z = \langle X, Y \rangle$ is distributed according to $\mathcal{D} \times \mathcal{E}$, where $X \in_{\mathcal{D}} S$ and $Y \in_{\mathcal{E}} T$ are independent random variables. We write

$$Z \in_{\mathcal{D} \times \mathcal{E}} S \times T$$

to indicate that Z is distributed according to the product distribution of $\mathcal{D}$ and $\mathcal{E}$. We often use the notation $X_1, \ldots, X_N \in_{\mathcal{U}} S$ to indicate that the random variables $X_1, \ldots, X_N$ are all uniformly distributed in S. The implicit assumption is that they are totally independent unless otherwise stated.

Definition (uniform distribution): We let $\mathcal{U}_n$ denote the uniform distribution on $\{0, 1\}^n$. ♣

Definition (correlation): Let $X \in_{\mathcal{U}} \{0, 1\}$ and let Y be a $\{0, 1\}$-valued random variable (not necessarily uniformly distributed or independent of X). The *correlation* of Y with X is $|\mathrm{E}[\overline{X} \cdot \overline{Y}]|$. ♣

Note that if X and Y are independent then the correlation is zero. Intuitively, the correlation of Y with X is a measure of how well Y predicts the value of X.

Asymptotics

Unless otherwise stated, there is implicit quantification over all $n \in \mathcal{N}$ in all statements involving the parameter n.

Asymptotics : Let both $k(n)$ and $\ell(n)$ be values in $\mathcal{N}$. We use the notation

- $k(n) = \mathcal{O}(\ell(n))$ if there is a constant $c > 0$ such that $k(n) \leq c \cdot \ell(n)$.

- $k(n) = \Omega(\ell(n))$ if there is a constant $c > 0$ such that $k(n) \geq c \cdot \ell(n)$.

- $k(n) = \ell(n)^{\mathcal{O}(1)}$ if there is a constant $c > 0$ such that $k(n) \leq \ell(n)^c$.

- $k(n) = \ell(n)^{\Omega(1)}$ if there is a constant $c > 0$ such that $k(n) \geq \ell(n)^c$.

- $k(n) = \ell(n^{\mathcal{O}(1)})$ if there is a constant $c > 0$ such that $k(n) \leq \ell(n^c)$.

- $k(n) = \ell(n^{\Omega(1)})$ if there is a constant $c > 0$ such that $k(n) \geq \ell(n^c)$.

Definition (non-negligible parameter): We say $k(n)$ is a *non-negligible parameter* if $k(n) = 1/n^{\Omega(1)}$ and if $k(n)$ is computable in time $n^{\mathcal{O}(1)}$ by a **TM**. ♣

Definition (polynomial parameter): We say $k(n)$ is a *polynomial parameter* if $k(n) = n^{\mathcal{O}(1)}$ and if $k(n)$ is computable in time $n^{\mathcal{O}(1)}$ by a **TM**. ♣

Ensembles

Function and probability ensembles are used to define primitives such as one-way functions and pseudorandom generators.

Definition (function ensemble): We let $f : \{0,1\}^{t(n)} \to \{0,1\}^{\ell(n)}$ denote a *function ensemble*, where $t(n)$ and $\ell(n)$ are polynomial parameters and where f with respect to n is a function mapping $\{0,1\}^{t(n)}$ to $\{0,1\}^{\ell(n)}$. If f is injective with respect to n then it is a *one-to-one function ensemble*. If f is injective with respect to n and $\ell(n) = t(n)$ then it is a *permutation ensemble*. We let $f : \{0,1\}^{t(n)} \times \{0,1\}^{\ell(n)} \to \{0,1\}^{m(n)}$ denote a function ensemble with two inputs. In this case, we sometimes consider f as being a function of the second input for a fixed value of the first input, in which case we write $f_x(y)$ in place of $f(x, y)$. ♣

Definition (P-time function ensemble): We say $f : \{0,1\}^{t(n)} \times \{0,1\}^{\ell(n)} \to \{0,1\}^{m(n)}$ is a $T(n)$-*time function ensemble* if f is a function ensemble and there is a **TM** such that, for all $x \in \{0,1\}^{t(n)}$, for all $y \in \{0,1\}^{\ell(n)}$, $f(x, y)$ is computable in time $T(n)$. We say f is a **P**-*time function ensemble* if $T(n) = n^{\mathcal{O}(1)}$. ♣

These definitions generalize in a natural way to functions with more than two inputs. Sometimes we describe functions that have a variable length inputs or outputs; in these cases we implicitly assume that the string is padded out with a special blank symbol to the appropriate length.

Definition (probability ensemble): We let $\mathcal{D}_n : \{0,1\}^{\ell(n)}$ denote a *probability ensemble*, where $\mathcal{D}_n$ is a probability distribution on $\{0,1\}^{\ell(n)}$. ♣

Definition (P-samplable probability ensemble): We let $\mathcal{D}_n : \{0,1\}^{r(n)} \to \{0,1\}^{\ell(n)}$ denote a probability ensemble on $\{0,1\}^{\ell(n)}$ that can be generated from a random string of length $r(n)$, i.e., there is a function ensemble $f : \{0,1\}^{r(n)} \to \{0,1\}^{\ell(n)}$ such that if $X \in_{\mathcal{U}} \{0,1\}^{r(n)}$ then $\mathcal{D}_n = f(X)$. We say $\mathcal{D}_n$ is $T(n)$-*samplable probability ensemble* if f is computable by a **TM** such that, for all $x \in \{0,1\}^{r(n)}$, $f(x)$ is computable in time $T(n)$. We say $\mathcal{D}_n$ is **P**-*samplable* if $T(n) = n^{\mathcal{O}(1)}$. ♣

We make the fundamental assumption that it is possible to produce independent, uniformly distributed random bits. A source of truly random bits is central to most of the definitions and constructions we describe in these lectures.

Definition (source of random bits): A *source of random bits* of length n is a sequence of n random bits distributed uniformly and independently of everything else, i.e., $X \in_{\mathcal{U}} \{0,1\}^n$. For simplicity, we assume that it takes $\mathcal{O}(n)$ time to produce the bits. ♣

In practice, bits that are supposedly random are produced by a variety of methods, including using the low order bits of the system clock, etc. A discussion of these methods are beyond the scope of these lectures.

We assume that a source of uniformly distributed random bits is the underlying source of randomness available. Random bits are often useful to efficiently solve problems that are difficult to solve deterministically.

Definition (randomized P-time function ensemble):
Let $f : \{0,1\}^n \times \{0,1\}^{r(n)} \to \{0,1\}^{\ell(n)}$ be a **P**-time function ensemble. We can view f as a *randomized* **P***-time function ensemble* that on input $x \in \{0,1\}^n$ produces the random output $f(x,Y)$, where $Y \in_{\mathcal{U}} \{0,1\}^{r(n)}$ is thought of as a random string that helps the computation. In this context, we present three possible definitions of a time bound.

- The worst-case time bound $T(n)$ of the randomized **P**-time function ensemble is the maximum over all $x \in \{0,1\}^n$ and $y \in \{0,1\}^{r(n)}$ the time to compute $f(x,y)$.

- The expected worst-case time bound $T'(n)$ of the randomized **P**-time function ensemble is the maximum over all $x \in \{0,1\}^n$ of the expected time to compute $f(x,Y)$ where $Y \in_{\mathcal{U}} \{0,1\}^{r(n)}$.

- Let $\mathcal{D}_n$ be a distribution on $\{0,1\}^n$ and let $X \in_{\mathcal{D}_n} \{0,1\}^n$. The expected average case time bound $T''(n)$ of the randomized **P**-time function ensemble with respect to $\mathcal{D}_n$ is the expected time to compute $f(X,Y)$, where $Y \in_{\mathcal{U}} \{0,1\}^{r(n)}$. ♣

Generally, we use the worst-case time bound for randomized **P**-time function ensembles.

Complexity Classes

Definition (language): Let $L : \{0,1\}^n \to \{0,1\}$ be a function ensemble. We can view L as a *language*, where, for all $x \in \{0,1\}^n$, $x \in L$ if

$L(x) = 1$ and $x \notin L$ if $L(x) = 0$.   ♣

In the following, we define various complexity classes with respect to a language L as just defined.

Definition (P): We say $L \in \mathbf{P}$ if L is a **P**-time function ensemble. ♣

Definition (NP): We say $L \in \mathbf{NP}$ if there is a **P**-time function ensemble $f : \{0,1\}^n \times \{0,1\}^{\ell(n)} \to \{0,1\}$ such that for all $x \in \{0,1\}^n$,

$$x \in L \quad \text{implies} \quad \Pr_{Y \in_u \{0,1\}^{\ell(n)}}[f(x,Y) = 1] > 0.$$
$$x \notin L \quad \text{implies} \quad \Pr_{Y \in_u \{0,1\}^{\ell(n)}}[f(x,Y) = 1] = 0.$$

♣

Definition (RP): We say $L \in \mathbf{RP}$ if there is a constant $\epsilon > 0$ and a **P**-time function ensemble $f : \{0,1\}^n \times \{0,1\}^{\ell(n)} \to \{0,1\}$ such that for all $x \in \{0,1\}^n$,

$$x \in L \quad \text{implies} \quad \Pr_{Y \in_u \{0,1\}^{\ell(n)}}[f(x,Y) = 1] \geq \epsilon.$$
$$x \notin L \quad \text{implies} \quad \Pr_{Y \in_u \{0,1\}^{\ell(n)}}[f(x,Y) = 1] = 0.$$

♣

Definition (BPP): We say $L \in \mathbf{BPP}$ if there are a pair of constants $\langle \epsilon, \epsilon' \rangle$ with $0 \leq \epsilon < \epsilon' \leq 1$ and a **P**-time function ensemble $f : \{0,1\}^n \times \{0,1\}^{\ell(n)} \to \{0,1\}$ such that for all $x \in \{0,1\}^n$,

$$x \in L \quad \text{implies} \quad \Pr_{Y \in_u \{0,1\}^{\ell(n)}}[f(x,Y) = 1] \geq \epsilon'.$$
$$x \notin L \quad \text{implies} \quad \Pr_{Y \in_u \{0,1\}^{\ell(n)}}[f(x,Y) = 1] \leq \epsilon.$$

♣

Definition (PP): We say $L \in \mathbf{PP}$ if there is a constant ϵ and a **P**-time function ensemble $f : \{0,1\}^n \times \{0,1\}^{\ell(n)} \to \{0,1\}$ such that for all $x \in \{0,1\}^n$,

$$x \in L \quad \text{implies} \quad \Pr_{Y \in_u \{0,1\}^{\ell(n)}}[f(x,Y) = 1] \geq \epsilon.$$
$$x \notin L \quad \text{implies} \quad \Pr_{Y \in_u \{0,1\}^{\ell(n)}}[f(x,Y) = 1] < \epsilon.$$

♣

Let L be a language in **NP** or **RP** as defined above and let $x \in \{0,1\}^n$. If $x \in L$ then there is a witness $y \in \{0,1\}^{\ell(n)}$ for which $f(x,y) = 1$. Such a y is called a "witness" to $x \in L$ because y can be used to certify that $x \in L$ simply by computing $f(x,y)$ and seeing that the answer is 1. This is a guarantee that $x \in L$ because if $x \notin L$ it is impossible that

$f(x, y) = 1$ for any $y \in \{0, 1\}^{\ell(n)}$. It is easy to see that **PP** and **BPP** languages do not necessarily have the witness property.

For both **RP** and **BPP** languages, membership in the language can be decided in the following sense by a randomized **P**-time function ensemble. If $L \in \mathbf{RP}$ then, for each $x \in \{0, 1\}^n$, if $x \in L$ then a fraction of at least ϵ of the $y \in \{0, 1\}^{\ell(n)}$ are witnesses. If we choose a random $y \in_{\mathcal{U}} \{0, 1\}^{\ell(n)}$, there is a chance of at least ϵ that y is a witness. If we randomly choose $y_1, \ldots, y_m \in_{\mathcal{U}} \{0, 1\}^{\ell(n)}$, the chance that none of them is a witness is at most $(1 - \epsilon)^m$. If we set $m = n$, then the probability that we don't find a witness is exponentially small in n. On input $x \in \{0, 1\}^n$, the randomized **P**-time function ensemble randomly chooses $y_1, \ldots, y_m$ at random and tests to see if there is some $i \in \{1, \ldots, m\}$ for which $f(x, y_i) = 1$. If the answer is yes, then x is classified as being in L, whereas if the answer is no then x is classified as not being in L. Note that the answer can never be incorrect when $x \notin L$, and the probability it is incorrect when $x \in L$ is exponentially small in n. If $L \in \mathbf{BPP}$, a similar randomized **P**-time function ensemble can be designed to test membership in L. In this case, the answer can be incorrect with probability exponentially small in n both when $x \in L$ and when $x \notin L$.

The same simple idea does not work for **NP** or **PP** languages. For example, let $L \in \mathbf{NP}$ and let $x \in \{0, 1\}^n$. If $x \in L$ then, although there is at least one witness $y \in \{0, 1\}^{\ell(n)}$ for x, the fraction of $y \in \{0, 1\}^{\ell(n)}$ that are witnesses can be exponentially small in $\ell(n)$. In this case, there is little chance that a witness can be found by randomly choosing $y \in_{\mathcal{U}} \{0, 1\}^{\ell(n)}$. The same type of reasoning applies to a **PP** language.

Perhaps the most celebrated problem in theoretical computer science is the question "Is $\mathbf{P} = \mathbf{NP}$?". Very little progress, at least as far as we know (given that it is often very hard to measure progress on a problem), has been made in resolving this question. We state two versions of the question.

- *Decision version:* For every **P**-time function ensemble $f : \{0, 1\}^n \times \{0, 1\}^{\ell(n)} \to \{0, 1\}$ is there a **P**-time function ensemble $g : \{0, 1\}^n \to \{0, 1\}$ such that for all $x \in \{0, 1\}^n$,

$$g(x) = \begin{cases} 1 & \text{if } f(x, y) = 1 \text{ for some } y \in \{0, 1\}^{\ell(n)} \\ 0 & \text{if } f(x, y) = 0 \text{ for all } y \in \{0, 1\}^{\ell(n)} \end{cases}$$

- *Search version:* For every **P**-time function ensemble $f : \{0, 1\}^n \times \{0, 1\}^{\ell(n)} \to \{0, 1\}$ is there a **P**-time function ensemble $g : \{0, 1\}^n \to$

$\{0,1\}^{\ell(n)}$ such that for all $x \in \{0,1\}^n$,

$$f(x,y) = 1 \text{ for some } y \in \{0,1\}^{\ell(n)} \text{ implies } f(x,g(x)) = 1.$$

Exercise 1 : Prove that the answer to the decision version of the **P = NP** question is yes if and only if the answer to the search version of the question is yes. ♠

Exercise 2 : One can generalize the definition of **BPP** to allow ϵ and ϵ' depend on n, i.e., $\epsilon(n)$ and $\epsilon'(n)$. Let **BPP**(gap(n)) denote the version of the **BPP** problem where gap(n) $= \epsilon'(n) - \epsilon(n)$. For any constant $1 > \epsilon > 0$ and for any constant $c > 0$, prove that

$$\mathbf{BPP}(1/n^c) = \mathbf{BPP}(\epsilon) = \mathbf{BPP}(1 - 1/2^{n^c}),$$

♠

Definition (P/poly): Let $L : \{0,1\}^n \to \{0,1\}$ be a function ensemble, and view L as a language. We say that $L \in \mathbf{P/poly}$ if there is a **P**-time function ensemble $f : \{0,1\}^n \times \{0,1\}^{\ell(n)} \to \{0,1\}$ and an advice string $y \in \{0,1\}^{\ell(n)}$ with the property that, for all $x \in \{0,1\}^n$,

$$x \in L \quad \text{implies } f(x,y) = 1.$$
$$x \notin L \quad \text{implies } f(x,y) = 0.$$

♣

We use the term "advice string" because, given the value of the advice string $y \in \{0,1\}^{\ell(n)}$, it is easy to decide membership in L for all $x \in \{0,1\}^n$. Note that if it is possible to compute the value of the advice string $y \in \{0,1\}^{\ell(n)}$ in $n^{\mathcal{O}(1)}$ time, then $L \in \mathbf{P}$. However, in general it may not be possible to compute the advice string in $n^{\mathcal{O}(1)}$ time. One way of thinking about a language $L \in \mathbf{P/poly}$ is that membership in L can be decided in $n^{\mathcal{O}(1)}$ time with the aid of a polynomial amount of extra advice for each input length.

Exercise 3 : Prove that $\mathbf{RP} \subseteq \mathbf{P/poly}$ and $\mathbf{BPP} \subseteq \mathbf{P/poly}$.

Hint : Use the idea discussed above for deciding membership in a **RP** or **BPP** language by a randomized **P**-time function ensemble. ♠

Useful Probability Facts and Inequalities

Markov inequality : Let $X \geq 0$ be a random variable such that $\mathrm{E}[X]$ is finite. Then, for all $\alpha > 0$, $\Pr[X \geq \alpha] \leq \frac{\mathrm{E}[X]}{\alpha}$. The following is the proof of this when the range of X is countable.

$$E[X] = \sum_{y \geq 0}^{< \alpha} y \cdot \Pr[X = y] + \sum_{y \geq \alpha} y \cdot \Pr[X = y].$$

Since y is always $\geq \alpha$ in the second summation, we have

$$\sum_{y \geq \alpha} y \cdot \Pr[X = y] \geq \alpha \cdot \sum_{y \geq \alpha} \Pr[X = y] = \alpha \cdot \Pr[X \geq \alpha].$$

Since $X \geq 0$, the first summation is always non-negative and thus the inequality follows.

Chebychev inequality : Let X be a random variable such that $E[X^2]$ is finite. Now X^2 is a random variable always ≥ 0. Then, from the Markov inequality, $\Pr[X^2 \geq \epsilon^2] \leq \frac{E[X^2]}{\epsilon^2}$. Equivalently $\Pr[|X| \geq \epsilon] \leq \frac{E[X^2]}{\epsilon^2}$. A more general form of this inequality is the following. Let f be a real-valued function, let I be a subinterval of the reals, let α_I be such that for all $x \notin I$, $f(x) \geq \alpha_I$. Let $\chi_I(x)$ be the indicator function for I, i.e., $\chi_I(x) = 1$ if $x \in I$ and $\chi_I(x) = 0$ if $x \notin I$. Then,

$$\alpha_I \cdot \Pr[X \notin I] + E[f(X) \cdot \chi_I(X)] \leq E[f(X)].$$

In typical applications of this bound, an upper bound is derived for $E[f(X)]$ and lower bounds are derived for $E[f(X) \cdot \chi_I(X)]$ and α_I, and these are used to derive an upper bound on the probability that the (typically bad) event $X \notin I$ occurs. Here are two example applications of this more general form:

- Let $k > 0$ be an integer and define $f(x) = x^{2k}$ and $I = (-\epsilon, -\epsilon)$. For all $x \notin I$, $f(x) \geq \epsilon^{2k}$, and $E[f(X)\chi_I(X)] \geq 0$, and thus

$$\Pr[|X| \geq \epsilon] \leq \frac{E[X^{2k}]}{\epsilon^{2k}}.$$

  This is often referred to as the kth-moment inequality. For $k = 1$, this is the special case we introduced above.

- Let $t \geq 0$ and μ be reals and define $f(x) = e^{(x-\mu)t}$ and $I = (-\infty, \mu)$. For all $x \notin I$, $f(x) \geq 1$, and $E[f(X) \cdot \chi_I(X)] \geq 0$, and thus

$$\Pr[X \geq \mu] \leq E[e^{(X-\mu)t}].$$

  In many applications, judicious choices of μ and t yield sharp probabilistic bounds.

Jensen inequality : Let X be a random variable such that $E[X^2]$ is finite. Then, $E[X^2] \geq E[X]^2$. This is because $(X - E[X])^2 \geq 0$ implies that $E[(X - E[X])^2] \geq 0$, and because $E[(X - E[X])^2] = E[X^2] - E[X]^2$. More generally,

$$E[f(X)] \geq f(E[X])$$

for any convex function f. A function f is convex if for all x, y, and for all $\alpha \in [0, 1]$, $f(\alpha x + (1 - \alpha)y) \leq \alpha f(x) + (1 - \alpha)f(y)$.

Chernoff bound : Let $X, X_1, \ldots, X_n$ be independent identically distributed $\{0, 1\}$-valued random variables. Let $p = \Pr[X = 1] < \frac{1}{2}$. Then, for all δ in the range $0 < \delta \leq p(1 - p)$,

$$\Pr\left[\left| \frac{1}{n} \sum_{i=1}^{n} X_i - p \right| \geq \delta \right] \leq 2e^{\frac{-\delta^2 n}{2p(1-p)}}.$$

Exercise 4 : Given X (not necessarily ≥ 0) such that $E[X] = \mu$ and $X \leq 2\mu$, give an upper bound on $\Pr[X < \frac{\mu}{2}]$ ♠

Exercise 5 : Let $X, X_1, \ldots, X_n$ be identically distributed and pairwise independent $\{0, 1\}$-valued random variables and let $p = \Pr[X = 1]$. Prove using the Chebychev inequality that:

$$\Pr\left[\left| \frac{1}{n} \sum_{i=1}^{n} X_i - p \right| \geq \delta \right] \leq \frac{p(1 - p)}{\delta^2 n}.$$

Pairwise independence means that for any pair $i, j \in \{1, \ldots n\}$, $i \neq j$ and for any pair $\alpha, \beta \in \{0, 1\}$,

$$\Pr[X_i = \alpha \wedge X_j = \beta] = \Pr[X_i = \alpha] \cdot \Pr[X_j = \beta].$$

Hint : Because the assumption is that the variables are pairwise independent, and for example not necessarily three-wise independent, you cannot use a Chernoff type bound to prove the result. Define $Y_i = X_i - p$ and $Z = \frac{1}{n} \sum_{i=1}^{n} Y_i$. Note that $E[Z^2] = \frac{1}{n^2} \sum_{i,j} E[Y_i \cdot Y_j]$. Because the variables are pairwise independent, the only non-zero terms in this sum are those for which $i = j$. ♠

Lecture 1

Overview

We describe a cryptosystem for sending a private message with the restriction that the message is at most as long as a previously established private key. We introduce the notion of a pseudorandom generator, and show how to use one to send messages much longer than the private key. In subsequent lectures, we show how to construct a pseudorandom generator from a one-way function. In this lecture, we informally introduce and give conjectured examples of one-way functions.

Private Key Cryptography

The first few lectures develop solutions to the following basic problem in cryptography: Alice and Bob are together now but soon they will be separated. When they are apart they can only communicate using a public line, i.e., a line that can be read by any outside party or adversary, but no information passing down the line can be modified in any way. While they are still together and isolated from the rest of the world, they choose a private random string, called the *private key*, that will be used to encrypt all future communication. After they are separated, when Alice wants to send a message to Bob, she first encrypts the message using the private key and then sends the encryption on a public line. Once the encryption is received by Bob, he decrypts it to recover the original message. The property desired from the encryption system is that an eavesdropper Eve is not able to deduce anything about the content of the messages from what is sent on the public line.

The private key is uniformly chosen from $\{0,1\}^n$. Because Eve is not privy to the exchange of the private key between Alice and Bob when they are still together, the knowledge about the private key just after it is selected is different for Alice and Bob than it is for Eve. Since Alice and Bob see the private key immediately once it is selected, from their perspective it is a fixed string $x \in \{0,1\}^n$. On the other hand, since Eve does not see the value of x, from Eve's perspective the private key is still a random variable $X \in_{\mathcal{U}} \{0,1\}^n$. This difference in knowledge is what allows Alice to send information on a public line to Bob without leaking any information to Eve.

One-time-pad Private Key Cryptosystem : Let m be a message that Alice wants to send to Bob. Consider the case when $\|m\| \leq n$ and m is the only message that Alice ever wants to send to Bob. Alice can send $y = m \oplus x$ on the public line. Upon receiving y, Bob can recover m by computing $x \oplus y = m$.

The question is what does an eavesdropper Eve see? The answer is random noise, i.e., for all messages $m \in \{0,1\}^n$, for all possible encryptions $y \in \{0,1\}^n$ of m,

$$\Pr_{X}[m \oplus X = y] = 1/2^n.$$

Restating this, from Eve's point of view (i.e., without knowledge of the private key), the distribution on strings y that Eve sees on the public line is the uniform distribution independent of the actual message m, and thus Eve receives no information about m from the encryption. This cryptosystem is perfectly secure in the information-theoretic sense.

Sending messages longer than the private key

The only problem is that Alice may want to send a message that is longer than the private key. Thus, a natural question to ask is what happens if the private key gets used up, i.e., if Alice and Bob did not initially agree on a private key that is long enough to encrypt all future messages.

Proposed Naive Solution : How about encrypting the message in blocks of n bits each as before? For instance, if $m = \langle m_1, m_2 \rangle$ and $\|m_1\| = \|m_2\| = n$ and $x \in \{0,1\}^n$ is the private key then one possibility is to send on the public line

$$\langle m_1 \oplus x, m_2 \oplus x \rangle.$$

Problems with naive solution : Since $(m_1 \oplus X) \oplus (m_2 \oplus X) = m_1 \oplus m_2$, Eve can learn a lot about the original message m. For example, if $m_1 = 0^n$ then Eve can compute m_2 from the two encryptions $m_1 \oplus X$ and $m_2 \oplus X$ independent of the value of the private key X.

The next few lectures develop a method for Alice and Bob to securely exchange messages on a public line so that the total length of all messages sent is greater than the length of the private key. Information-theoretically, this is an impossible task, i.e., it can be shown that if the total length of the messages is greater than the length of the private key then, no matter what encryption system is used, Eve provably has some information about the content of the messages from the encryptions sent on the public line.

However, Eve may not have enough computational resources (e.g., time) to be able to compute any revealing information about the content of the sent messages. The idea is to exploit the computational limitations of Eve. Intuitively, what Alice and Bob want to do is to encrypt very long messages using a short random private key in such a way that the encryptions are indistinguishable from truly random noise to any eavesdropper with reasonable computational limits. At the heart of the encryption system we use to implement these ideas is a pseudorandom generator.

Definition (distinguishing probability): Let $A : \{0,1\}^n \to \{0,1\}$ be a function ensemble and let X and Y be random variables distributed on $\{0,1\}^n$. The *distinguishing probability* of A for X and Y is

$$\delta(n) = |\Pr_X[A(X) = 1] - \Pr_Y[A(Y) = 1]|.$$

♣

Definition (pseudorandom generator [very informal]): Let $g : \{0,1\}^n \to \{0,1\}^{\ell(n)}$ be a **P**-time function ensemble, where $\ell(n) > n$, and let $X \in_{\mathcal{U}} \{0,1\}^n$. We say that g is a *pseudorandom generator* if $g(X)$ "looks like" a truly random string $Z \in_{\mathcal{U}} \{0,1\}^{\ell(n)}$. Intuitively, $g(X)$ "looks like" Z means that, for any **P**-time function ensemble $A : \{0,1\}^{\ell(n)} \to \{0,1\}$, the distinguishing probability of A for $g(X)$ and Z is very small. ♣

Given a pseudorandom generator g, Alice can easily encrypt a message $m \in \{0,1\}^{\ell(n)}$ using private key $x \in \{0,1\}^n$ by computing $g(x)$ and sending on the public line $y = g(x) \oplus m$. Since Bob knows x, he can recover m by computing $g(x) \oplus y$.

Recall that from Eve's point of view the unknown private key is a random variable $X \in_{\mathcal{U}} \{0,1\}^n$. Intuitively, to $n^{\mathcal{O}(1)}$ time bounded Eve, $g(X)$ looks just like $Z \in_{\mathcal{U}} \{0,1\}^{\ell(n)}$. Thus, for any message m, Eve should not be able to tell the difference between $m \oplus g(X)$ and $m \oplus Z$. Since Eve can't distinguish $m \oplus g(X)$ and $m \oplus Z$, and $m \oplus Z$ gives no information about m, it follows that Eve computationally has no more information about m after seeing $m \oplus g(X)$ than before. Thus, the above cryptosystem is a computationally secure cryptosystem if g is a pseudorandom generator. We later formalize and quantify these informal notions.

It is not known if pseudorandom generators exist, but it is easy to show that if they do exist then $\mathbf{P} \neq \mathbf{NP}$, i.e., see Exercise 8 below.

The conditions for a pseudorandom generator are rather stringent, and it is not easy to come up with a natural candidate. On the other hand, there seem to be a variety of natural examples of another basic primitive; the one-way function.

Definition (one-way function [very informal]): Let $f : \{0,1\}^n \to \{0,1\}^{\ell(n)}$ be a **P**-time function ensemble, and let $X \in_{\mathcal{U}} \{0,1\}^n$. We say that f is a *one-way function* if $f(X)$ is "hard to invert on average". Intuitively, "hard to invert on average" means that, for any **P**-time function ensemble $A : \{0,1\}^{\ell(n)} \to \{0,1\}^n$, the probability that $A(f(X))$ is an inverse of $f(X)$ is small.   ♣

It has not been proven that one-way functions exist, but there are plenty of natural candidates that might eventually be proven to be one-way functions. On the other hand, if $\mathbf{P} = \mathbf{NP}$ then there are no one-way functions.

Exercise 6 : Show that $\mathbf{P} = \mathbf{NP}$ implies there are no one-way functions.

Hint : Let $f : \{0,1\}^n \to \{0,1\}^{\ell(n)}$ be any **P**-time function ensemble. Let $M : \{0,1\}^{\ell(n)} \times \{0,1\}^n \to \{0,1\}$ be a **P**-time function ensemble with the property that $M(y,x) = 1$ if $f(x) = y$ and $M(y,x) = 0$ otherwise. Use Exercise 1 (page 10) to show that if $\mathbf{P} = \mathbf{NP}$ then this implies there is a **P**-time function ensemble $N : \{0,1\}^{\ell(n)} \to \{0,1\}$ with the property that, for all $x \in \{0,1\}^n$, $f(N(f(x))) = f(x)$, i.e., N on input $f(x)$ produces an inverse of $f(x)$.   ♠

Note : We do not know whether or not the converse of this exercise is true, i.e., it is not known whether or not $\mathbf{P} \neq \mathbf{NP}$ implies that there are one-way functions. The difficulty is that a one-way function is hard to invert with respect to a **P**-samplable distribution, but a proof that there is a function in **NP** that is not in **P** does not necessarily imply that there is a **P**-samplable distribution on which it is hard to invert this function.

————————∞————————

One of the main results we develop in these lectures is the construction of a pseudorandom generator from any one-way function. The constructions have the property that if there is a **TM** that can distinguish the output of the pseudorandom generator from a truly random string then we can use this **TM** to construct another **TM** with close to the same running time that inverts the one-way function.

Examples of Conjectured one-way functions

Here are some natural examples that may eventually be proven to be one-way functions. Plenty of others can be found in the literature. In the following, p and q are primes of length n.

Factoring problem : Define $f(p, q) = pq$. It is possible to compute pq given p and q in $n^{\mathcal{O}(1)}$ time. However, there is no known **P**-time function ensemble that on input pq can produce p and q on average for randomly chosen pairs of primes $\langle p, q \rangle$

Discrete log problem : Let g be a generator of $\mathcal{Z}_p^*$, i.e., for all $y \in \mathcal{Z}_p^*$, there is a unique $x \in \mathcal{Z}_{p-1}$ such that $g^x = y \bmod p$. Given p, g and $x \in \mathcal{Z}_{p-1}$, define $f(p, g, x) = \langle p, g, g^x \bmod p \rangle$. We view p and g as public inputs and x as the private input. It is possible to compute $g^x \bmod p$ given p, g and x in $n^{\mathcal{O}(1)}$ time. The discrete log function is a permutation, i.e., the unique inverse of $f(p, g, x)$ is $\langle p, g, x \rangle$. The values of p and g are not necessarily chosen randomly. The prime p is selected to have special properties which seem in practice to make the discrete log function hard to invert. An example of such a property is that p is selected so that that $p - 1$ has some fairly large prime divisors. For a large class of primes p and generators g there is no known **P**-time function ensemble that on input p, g and $g^x \bmod p$ can produce x on average for $x \in_{\mathcal{U}} \mathcal{Z}_{p-1}$.

Root extraction problem : Given a pair of primes p and q, a value $e \in \mathcal{Z}_{pq}$ relatively prime to $(p-1)(q-1)$, and $y \in \mathcal{Z}_{pq}$, define $f(p, q, e, y) = \langle pq, e, y^e \bmod pq \rangle$. We view the exponent e as a public input and p, q and y as private inputs. It is possible to compute $y^e \bmod pq$ given pq, e and y in $n^{\mathcal{O}(1)}$ time. For fixed values for p, q and e, the function is a permutation as a function of y. To make the inversion problem hard, it is important that the factorization of the modulus is not part of the output, because given the factorization an inverse can be found in $n^{\mathcal{O}(1)}$ time. This problem is commonly known as the RSA function.

———————∞———————

The value of the exponent e need not necessarily be chosen randomly. For example, the Rabin function sets $e = 2$, and then the problem is to extract square roots, and this still seems to be a hard problem on average. In this case, for fixed values of p, q, $e = 2$, the function is 4-to-1 as a function of y.

For either of these versions, there is no known **P**-time function ensemble that on input pq, e and $y^e \bmod pq$ can produce a $y' \in \mathcal{Z}_{pq}$ such that

$y'^e = y^e \bmod pq$ when p and q are randomly chosen according to a distribution for which factoring is hard and $y \in_{\mathcal{U}} \mathcal{Z}_{pq}$. As we show on page 147, there is a strong connection between the Rabin version of this problem when $e = 2$ and the factoring problem.

Subset sum problem : Let $a \in \{0,1\}^n$ and $b \in \{0,1\}^{n \times n}$. Given a and b, define $f(a,b) = \langle \sum_{i=1}^n a_i \cdot b_i, b \rangle$, where $a_i \in \{0,1\}$ and b_i is an n-bit integer in this expression, and where the sum is over the integers. It is possible to compute $\sum_{i=1}^n a_i \cdot b_i$ given a and b in $n^{\mathcal{O}(1)}$ time. However, there is no known **P**-time function ensemble that on input $\sum_{i=1}^n a_i \cdot b_i$ and b can produce $a' \in \{0,1\}^n$ such that $\sum_{i=1}^n a_i' \cdot b_i = \sum_{i=1}^n a_i \cdot b_i$ on average when $a \in_{\mathcal{U}} \{0,1\}^n$ and $b \in_{\mathcal{U}} \{0,1\}^{n \times n}$.

Exercise 7 : Let $A \in_{\mathcal{U}} \{0,1\}^n$ and let $B \in_{\mathcal{U}} \{0,1\}^{n \times (n+1)}$. Prove the probability that

$$f(A, B) = \langle \sum_{i=1}^n A_i \cdot B_i, B \rangle$$

has a unique inverse is lower bounded by a constant strictly greater than zero independent of n.   ♠

These lectures develop general techniques for constructing cryptographic protocols based on one-way functions. However, what is sadly lacking in the field are reductions between specific conjectured one-way functions, e.g., reductions of the form "factoring is hard iff subset sum is hard". Even more specific reductions between instances of the same problem are in general not known, e.g., reductions of the form "discrete log mod p is hard iff discrete log mod q is hard", where p and q are somehow related to one another. One exception where this specific kind of reduction is known is for the Subset sum problem, as described in the references to Lecture 6 on page 204.

Protocol and Adversary Resources

For any cryptographic primitive, such as a pseudorandom generator g, there are two important parts to the definition:

(1) There is a *party* (or set of parties) that compute the cryptographic primitive in a reasonable amount of time, e.g., a **P**-time function ensemble $g : \{0,1\}^n \to \{0,1\}^{\ell(n)}$ that is a pseudorandom generator.

(2) The cryptographic primitive is secure against *adversaries* that run in a reasonable amount of time, e.g., for every **P**-time function

ensemble $A : \{0,1\}^{\ell(n)} \to \{0,1\}$, the distinguishing probability of A for $g(X)$ and $Z \in_{\mathcal{U}} \{0,1\}^{\ell(n)}$ is very small, where $X \in_{\mathcal{U}} \{0,1\}^n$.

Definition (party [informal]): A *party* is a randomized **P**-time function ensemble. The time bound is worst case. In some cases, a party interacts in the protocol with one or more other parties. In these cases, we still think of a party as being a randomized **P**-time function ensemble, but with the ability to interact with other randomized **P**-time function ensembles. ♣

Definition (adversary [informal]): An *adversary* is typically a **TM** or a randomized **TM**. The time bound is worst case. ♣

Definition (success probability [informal]): The *success probability* of an adversary is particular to the cryptographic primitive in question, and thus its definition is given within the definition of the primitive in each case. As examples, for a one-way function the success probability is the probability that the adversary inverts the function, and for a pseudorandom generator the success probability is the distinguishing probability. ♣

Definition (time-success ratio [informal]): How well an adversary breaks a cryptographic primitive is measured in terms of its time bound $T(n)$ and its success probability $\delta(n)$ on inputs parameterized by n. We combine these two quantities into a single *time-success ratio* $T(n)/\delta(n)$. ♣

Definition (one-way function [informal]): Let $f : \{0,1\}^n \to \{0,1\}^{\ell(n)}$ be a **P**-time function ensemble and let $X \in_{\mathcal{U}} \{0,1\}^n$. Let $A : \{0,1\}^{\ell(n)} \to \{0,1\}^n$ be an adversary. The success probability of A for f is

$$\delta(n) = \Pr_X[f(A(f(X))) = f(X)].$$

Then, f is a *one-way function* if there is no adversary for f with time-success ratio $n^{\mathcal{O}(1)}$. ♣

Note that the parameter n for the adversary A is not the length of the input to A, which is $\ell(n)$, but rather the length of the input to f that generated the input to A.

Definition (pseudorandom generator [informal]): Let $g : \{0,1\}^n \to$ $\{0,1\}^{\ell(n)}$ be a **P**-time function ensemble, where $\ell(n) > n$, let $X \in_{\mathcal{U}}$ $\{0,1\}^n$ and let $Z \in_{\mathcal{U}} \{0,1\}^{\ell(n)}$. Let $A : \{0,1\}^{\ell(n)} \to \{0,1\}$ be an adversary. The success probability of A for g is the distinguishing probability

$$\delta(n) = |\Pr_X[A(g(X)) = 1] - \Pr_Z[A(Z) = 1]|.$$

Then, g is a *pseudorandom generator* if there is no adversary for g with time-success ratio $n^{\mathcal{O}(1)}$.   ♣

Exercise 8 : Show that $\mathbf{P} = \mathbf{NP}$ implies there are no pseudorandom generators. In particular, show that $\mathbf{P} = \mathbf{NP}$ implies that for any **P**-time function ensemble $g : \{0,1\}^n \to \{0,1\}^{\ell(n)}$, with $\ell(n) > n$, there is a **P**-time function ensemble $A : \{0,1\}^{\ell(n)} \to \{0,1\}$ such that the success probability $\delta(n)$ of A for g is as large as possible, i.e., $\delta(n) = 1 - 2^{-\ell(n)+n}$.   ♠

Lecture 2

Overview

We discuss security issues associated with the computing environment of a party, and define the security parameter of a primitive based on this discussion. Adversaries that try to break primitives are introduced, together with the notion of time-success ratio, and the security of a primitive is defined. Definitions of one-way functions and one-way permutations are given, and cryptographic reduction is defined.

Introduction

The definition of a primitive includes the description of the interaction between the parties that implement the primitive and the allowable behavior of adversaries trying to break it. As an informal example, a function f is said to be a one-way function if it is easy to compute by a party but hard to invert for any $n^{\mathcal{O}(1)}$ time bounded adversary. The bulk of these lectures are devoted to reductions between primitives, e.g., a reduction from a one-way function to a pseudorandom generator. Examples of other primitives considered are pseudorandom function generators, pseudorandom invertible permutation generators, universal one-way hash functions, digital signatures, bit commitment, etc.

Descriptions of primitives and reductions are parameterized by n. An instance of a primitive is actually a family of instances, one for each $n \in \mathcal{N}$. For example, a one-way function $f : \{0,1\}^n \to \{0,1\}^{\ell(n)}$ is a collection of functions, one function $f : \{0,1\}^n \to \{0,1\}^{\ell(n)}$ for each $n \in \mathcal{N}$. Similarly, a reduction is actually a family of reductions, one reduction for each $n \in \mathcal{N}$. The analysis we give of the security preserving properties of reductions is asymptotic. In practice a reduction is used for one fixed value of n. As we stress in greater detail later, quantitative statements about the security preserving properties of reductions for fixed values of n can be easily derived from the asymptotic results.

Parties

A party is a randomized **P**-time function ensemble with a few additional security properties. A party may use memory with two levels of security.

- *Public memory:* The only security property required of this type of memory is that it cannot be changed by outside parties or by an adversary. However, it may be possible for an outside party or an adversary to read the contents of this memory. Thus, this memory is write protected but not read protected.

- *Private memory:* This memory cannot be accessed in any way by an outside party or adversary. This memory is both write protected and read protected.

In practice, because of the much more stringent security requirements, it is much more expensive to implement private memory than it is to implement public memory, and thus the amount of private memory required by a party to enact a primitive is a crucial security concern.

Similarly, a party may use computational devices with different levels of security.

- *Public computational device:* The only security property required of this computational device is that the results of the computation cannot be changed by outside parties or by an adversary, but it may be possible for them to see the internal states of the computation.

- *Private computational device:* No portion of this computational device can be observed by an outside party or by an adversary while computation is being performed. Typically, computation in this device depends on the contents of the private memory and sometimes portions of the output are to be stored in the private memory. Thus, there is a read and write protected data path between the private memory and this device.

Although the size of the private computational device is also a crucial security concern, it is perhaps less so than the size of the private memory. This is because the private memory must be protected at all points in time, whereas the private computational device need only be protected during actual computation if all internal state information is destroyed at the end of the computation. As we indicate below for specific cryptographic primitives, typically the amount of time needed to perform the private computations is small compared to the amount of time information stays in the private memory.

The parties we describe are usually randomized, i.e., they use sources of random bits. We distinguish between two kinds of sources of random bits, public random bits and private random bits. Public random bits generated by a party cannot be changed by any outside party or adversary, whereas private random bits have the additional property that they cannot be read by any outside party or adversary. Typically, private random bits are stored in the private memory. Thus, there is a direct connection from the source of private random bits into the private memory that is both read and write protected.

In many cases, a primitive requires two or more parties to communicate with one another. It is important to distinguish the types of communication channel they use for these exchanges of information. We distinguish between three types of communication:

- *Private line:* This is a line that connects a particular sending party to a particular receiving party, and is used by the sending party to send information directly to the receiving party. It is impossible for other parties to tamper with information sent down the line, and thus the receiving party has a guarantee that all information sent along the line is from the sending party. In addition, no other party or adversary can read any information sent on the line.

- *Public line:* This is a line that connects a particular sending party to a particular receiving party, and is used by the sending party to send information directly to the receiving party. It is impossible for other parties to tamper with information sent down the line, and thus the receiving party has a guarantee that all information sent along the line is from the sending party. However, all parties and adversaries can read all information sent on the line.

- *Public network:* This is a network that connects a group of parties for shared communications. This is a connectionless oriented type of communication, i.e., a party receiving information cannot determine directly where the information came from. In addition, information sent on a public network can be read by any party or adversary, and can be deleted or tampered with by any party or adversary.

It is clear that a private line is the hardest and most costly to implement and that a public network is the easiest and cheapest. This is true not only with respect to fixed cost, but also with respect to the price per bit communicated, e.g., the private line need only be protected while the line is in use. For this reason, when we implement a primitive we would always like to use the weakest type of communication line possible. If a private line is used at all, it is typically only used in the initial phase (when the two parties are perhaps in the same location), and thereafter all communication uses either a public line or public network.

Security Parameter

Perhaps the most important property of an instance of a primitive is the level of security it achieves. The exact definition of the security of a primitive depends on the particular primitive. In all cases the amount of security achieved by a particular instance is parameterized by the size of the private memory used by the instance. This certainly makes sense when comparing the cost of implementing the private memory versus the cost of implementing either the public memory or the public computational device, since the private memory is much harder and more costly to implement. It may seem that it is just as expensive to implement the private computational device as it is to implement the private memory. It turns out that typically this is not the case, i.e., in general the private computational device need only be protected for short periods of time compared to the amount of time the contents of the private memory must remain hidden. For most primitives, if the information stored in the private memory is released then all security is lost.

A typical example is that one party wants to send information privately to another party based on a pseudorandom generator. The random input to the generator is produced privately by one of the two parties and sent using a private line to the other party in an initial phase, and thereafter all communication uses a public line. In this case, the random input to the generator is stored in the private memory by both parties and must be protected for as long as the messages sent are to be kept private from any adversary. On the other hand, although the output of the generator is computed in the private computational device, the computation time is typically very small compared to the amount of time the messages sent are to be kept private. Once the output is produced it can be stored in public memory, the information stored in the private computational device can be immediately destroyed, and the private computational device no longer needs to be protected.

Definition (the security parameter): The *security parameter* of an instance of a primitive is the size of the private memory $s(n)$ associated with the nth instance.   ♣

A typical example is that the security parameter of a one-way function is the length of its input. Although the primary resource considered in these lectures to achieve a certain level of security is the amount of private memory used, the other resources are also important and emphasis should be put on minimizing these as well. Examples of these other resources are the running time of the parties to execute the instance, the number of random bits they use, the size of the private computational device, etc.

Adversaries and Security

An adversary tries to break an instance of a primitive. We consider two types of adversaries, uniform and non-uniform. A uniform adversary is a function ensemble that can be computed by a **TM** or a randomized **TM**, whereas a non-uniform adversary is a function ensemble that can be computed by a circuit family.

Definition (circuit family): A *circuit family* A is a family of circuits, one circuit A_n (with "and", "or" and "not" gates and $\{0,1\}$-valued inputs) for each value of n. The time bound $T(n)$ of A is the size of A_n, i.e., the number of gates and wires in A_n. The circuit may also have access to a source of random bits. ♣

The time bound $T(n)$ in the definition of a circuit family is an upper bound on the time to compute an output of A_n given the description of A_n and an input. The time bound does not include the time for finding the description of A_n, which could be exponentially large in the size of this description. Because of this, a circuit family is less desirable than a **TM** with the same time bound.

Definition (adversary): An *adversary* is a function ensemble that can be computed by a **TM**. A non-uniform adversary is a function ensemble that can be computed by a circuit family. ♣

Intuitively, the security of an instance of a primitive measures the computational resources needed by any adversary to break it. There are two natural computational resources we consider: The total time $T(n)$ the adversary runs and the success probability $\delta(n)$ of the adversary. The definition of the success probability of an adversary is primitive dependent. Generally, it measures the average success of the adversary for a randomly chosen input. We adopt the convention that the running time of an adversary is worst case. It turns out to be convenient to use a single time-success ratio to measure how well an adversary breaks an instance of a primitive.

Definition (time-success ratio): The *time-success ratio* of an adversary A for an instance f of a primitive is defined as $\mathbf{R}(\mathbf{s}(n)) = T(n)/\delta(n)$, where $T(n)$ is the worst case time bound of A, $\delta(n)$ is the success probability of A for f and $\mathbf{s}(n)$ is the security parameter of f. ♣

The time-success ratio of A for f is parameterized by the size of the private memory $s(n)$ used by f, and not by n. The reason for this single measure is simplicity, and because of the following generic example.

Example : Let $f : \{0,1\}^n \to \{0,1\}^{\ell(n)}$ be a one-way function. Let $A : \{0,1\}^{\ell(n)} \to \{0,1\}^n$ be an adversary for f with run time $T(n)$ and success probability $\delta(n)$. Let $p(n) < T(n)$ and let adversary $A'(y)$ work as follows: With probability $p(n)/T(n)$ run adversary $A(y)$ and with probability $1 - p(n)/T(n)$ do nothing. If we assume that $p(n)/T(n)$ can be computed in time $p(n)$ then the expected running time $T'(n)$ of $A'(y)$ is $\mathcal{O}(p(n))$, whereas the success probability of A' is $\delta'(n) = \delta(n)p(n)/T(n)$. If we use the expected run time of A' in the definition of its time-success ratio then the ratio is $T'(n)/\delta'(n) = \mathcal{O}(T(n)/\delta(n))$. Thus, we can reduce the expected run time of the original adversary A at the expense of a proportional decrease in the success probability, but the time-success ratio remains basically the same.

———————∞———————

In the definition of the time-success ratio, the run time of the adversary is worst case. One may question this convention, because it may seem overly restrictive to use worst case instead of average case run time. The reason for using worst case run time is because it is simpler, both in the definitions and in the proofs showing security preserving properties of reductions from one primitive to another. The following exercise shows that there is not much loss in generality in our choice.

Exercise 9 : Let $f : \{0,1\}^n \to \{0,1\}^{\ell(n)}$ be a one-way function and let $X \in_{\mathcal{U}} \{0,1\}^n$. Let $A : \{0,1\}^{\ell(n)} \to \{0,1\}^n$ be a deterministic adversary for f. The average case time-success ratio of A is

$$1/\mathrm{E}_X[\chi_A(f(X))/T_A(f(X))],$$

where $\chi_A(f(x)) = 1$ if A is successful in inverting $f(x)$ and $\chi_A(f(x)) = 0$ if A is unsuccessful in inverting $f(x)$, and where $T_A(f(x))$ is the running time of A on input $f(x)$. Show there is an adversary A' with worst case time-success ratio at most n times the average case time-success ratio of A. ♠

Definition (security): An instance f of a primitive is $\mathbf{S}(s(n))$-*secure* if every adversary A for f has time-success ratio $\mathbf{R}(s(n)) \geq \mathbf{S}(s(n))$. ♣

Let $p(n)$ be the run time of the parties implementing an instance f. In general, f is not secure in a useful sense if there is an adversary A with time-success ratio not much larger than $p(n)$, i.e., an instance is only useful if it is harder to break than it is to implement.

In this monograph, we are careful to quantify the amount of security achieved by an instance of a primitive, and when we describe a reduction of one primitive to another we are careful to quantify how much of the security of the first instance is transferred to the second. This approach is not always followed in the cryptographic literature. For comparison purposes, we give the definition of security commonly found in the literature.

Commonly used notion of security : An instance f is secure if there is no adversary for f with time-success ratio $s(n)^{\mathcal{O}(1)}$.

──────────∞──────────

This commonly used notion is not flexible enough to quantify security for many applications. For example, one-way functions are often conjectured to be very secure, e.g., $2^{s(n)^c}$-secure for some constant $c \leq 1$. At the other end of the spectrum, in some applications a low level of security for a one-way function may be enough, e.g., $s(n)^{100}$-secure. Neither end of this security spectrum can be quantified with the commonly used notion. Furthermore, the commonly used notion cannot be used to quantify the amount of security a reduction preserves.

═══════════════════════════════════════

Definitions of one-way functions and one-way permutations

We now give the formal definitions of a one-way function and a one-way permutation when the entire input is considered to be private.

Definition (one-way function): Let $f : \{0,1\}^n \rightarrow \{0,1\}^{\ell(n)}$ be a P-time function ensemble with security parameter $\|x\| = n$. Let $X \in_{\mathcal{U}} \{0,1\}^n$. The success probability (inverting probability) of adversary A for f is

$$\delta(n) = \Pr_X[f(A(f(X))) = f(X)].$$

Then, f is a $S(n)$-*secure one-way function* if every adversary has time-success ratio at least $S(n)$. ♣

The use of a one-way function that justifies parameterizing security by $\|x\|$ is the following. A party produces x using its private random source and stores x in its private memory. The party uses its private computational device to compute $f(x)$ from x, and stores this result in public memory (and immediately destroys any partial results still left in the private computational device). An adversary, which has read access to $f(x)$ in the public memory, tries to produce an inverse of $f(x)$. In a typical application, the protocol just described is a subprotocol embedded in a much more involved overall protocol and the party keeps x stored in the private memory for long periods of time.

The uniform distribution on X in the definition can be generalized to be any **P**-samplable distribution. This same remark holds for most of the definitions made with respect to the uniform distribution.

Definition (one-way permutation): Exactly the same as the definition of a one-way function, except that $\ell(n) = n$ and f as a function of $x \in \{0, 1\}^n$ is a permutation, i.e., f is a permutation ensemble. ♣

Functions with public input

Most of the primitives we introduce involve computing one or more **P**-time function ensembles by parties. In the traditional definitions of cryptographic functions, e.g., one-way functions and pseudorandom generators, the entire input to the function is assumed to be private. Since these functions are computed by parties and parties have two different levels of memory protection, it is natural and in many cases useful to distinguish parts of the input as being either private or public. The primary reason for distinguishing the two types of inputs is to parameterize security in an appropriate way, i.e., solely by the length of the private part of the input.

Public input to a function : Some functions have both private and public inputs. The security parameter of a function is the length of its private input. When we define primitives such as a one-way functions, we specify which inputs are kept private and which inputs are public, i.e., known to both parties and any adversary.

––––––––––∞––––––––––

When a party is computing a function, the private part of the input is stored in the private memory and the public part is stored in the public memory. Typically, both the private and public parts of the inputs are random strings; the party produces these random bits using the private and public sources of random bits, respectively. Although the public part of the input is available to all outside parties and adversaries, it turns out that these bits often play a crucial role in ensuring that a particular instance of a primitive is secure.

Definition (one-way function with a public input): Let $f : \{0, 1\}^{p(n)} \times \{0, 1\}^n \to \{0, 1\}^{\ell(n)}$ be a **P**-time function ensemble where the first input is public and the second private, and thus the security parameter is n. Let $Y \in_{\mathcal{U}} \{0,1\}^{p(n)}$ and $X \in_{\mathcal{U}} \{0,1\}^n$. The success probability of adversary A for f is

$$\delta(n) = \Pr_{X,Y}[f_Y(A(f_Y(X), Y)) = f_Y(X)].$$

Then, f is a $\mathbf{S}(n)$-*secure one-way function* if every adversary has time-success ratio at least $\mathbf{S}(n)$. ♣

Definition (one-way permutation with public input): Exactly the same as the definition of a one-way function with public input, except that $\ell(n) = n$ and for every fixed $y \in \{0,1\}^{p(n)}$, f_y as a function of $x \in \{0,1\}^n$ is a permutation. ♣

To exemplify the difference between the traditional definition of a one-way function with just private input and the definition introduced here with both private and public inputs, consider the Subset sum problem (page 18). A one-way function based on the difficulty of this problem can be defined in two ways; where the entire input is considered to be private and where the input is broken into private and public parts. Let $a \in \{0,1\}^n$, $b \in \{0,1\}^{n \times n}$, and recall that

$$f(a,b) = \langle \sum_{i=1}^n a_i \cdot b_i, b \rangle.$$

In the first definition, where the entire input is considered to be private, the security parameter is $\|a\| + \|b\| = n + n^2$, even though b is available to the adversary when trying to produce an inverse. In the second definition, b is considered to be public, and the security parameter is $\|a\| = n$. In both cases the security is based on exactly the same thing, i.e., when a and b are chosen uniformly then, given $\alpha = \sum_{i=1}^n a_i \cdot b_i$ and b, there is no fast adversary that on average can find a $a' \in \{0,1\}^n$ such that $\sum_{i=1}^n a_i' \cdot b_i = \alpha$. The only difference is how the security parameter is defined. The second definition, where the security is parameterized solely by what is kept private from the adversary, makes the most sense.

Cryptographic Reductions

Most of the results we present show how to reduce one cryptographic primitive to another. Examples of the reductions we present are:

- From a weak one-way function to a one-way function.

- From a one-way function to a pseudorandom generator.

- From a pseudorandom generator to a pseudorandom function generator.

- From a pseudorandom function generator to a pseudorandom invertible permutation generator.

We define two types of reductions, uniform and non-uniform. Most of the reductions we describe are the more desirable uniform type. We discuss non-uniform reductions later. We use the following definition in our description of a reduction.

Definition (oracle adversary): An *oracle adversary* is an adversary S that is not fully specified in the sense that S, in the course of its computations, interactively makes queries (hereafter described as *oracle queries*) to, and receives corresponding outputs from, an adversary that is not part of the description of S. We let S^A denote the fully specified adversary described by S making oracle queries to adversary A. The run time of S^A includes the time for computing A in the oracle calls to A. ♣

The following is an example of an oracle adversary S that makes one oracle query to an adversary $A : \{0,1\}^n \rightarrow \{0,1\}$.

---

oracle adversary S^A on input $x \in \{0,1\}^n$:

Randomly choose $\alpha \in_{\mathcal{U}} \{0,1\}$.

If $A(x) = 1$ then output α

Else output 1.

For example, if $A(x) = 1$ independent of x then $S^A(x) \in_{\mathcal{U}} \{0,1\}$, whereas if $A(x) = 0$ independent of x then $S^A(x) = 1$. Although the running time of S is not defined, the running time of S^A is defined. Also, if A is a **TM** then so is S^A.

Definition (P-time oracle adversary): An **P**-*time oracle adversary* is an oracle adversary P with the property that if M is a **P**-time function ensemble then P^M is a **P**-time function ensemble.   ♣

We now define what it means to reduce one primitive to another in the special case when both primitives can be described by a **P**-time function ensemble. (This is the case, for example, in the reduction of a one-way function to a pseudorandom generator.)

Definition (uniform reduction): We say that there is a *uniform reduction* from primitive 1 to primitive 2 if there is an **P**-time oracle adversary P and an oracle adversary S with the following properties.

- *Construction:* Given any **P**-time function ensemble f that is an instance of primitive 1 with security parameter $s(n)$, P^f is a **P**-time function ensemble g that is an instance of primitive 2 with security parameter $s'(n)$.

- *Guarantee:* Given any adversary A for g with time-success ratio $\mathbf{R}'(\mathbf{s}'(n))$, S^A is an adversary for f with time-success ratio $\mathbf{R}(\mathbf{s}(n))$. ♣

The construction is invariably described first and the guarantee is usually described within a proof. For the proof of the guarantee, we assume the existence of an adversary A for g and prove that S^A is an adversary for f. For simplicity, we always assume that A is a deterministic **TM**. On the other hand, it is more reasonable to assume that A is a randomized **TM**, and in almost all of our reductions it turns out that S^A is a randomized **TM** even if A is deterministic. It can be checked that this seeming disparity is not crucial, i.e., the analysis of all reductions we give would basically be the same if we assumed A was a randomized **TM**.

Security Preserving Reductions

A crucial property of a reduction from f to g is how much security is maintained by the reduction, i.e., a reduction should inject as much of the security of f as possible into g. To measure this fairly, we compare the time-success ratios when both f and g use the same amount of private information. We would like $\mathbf{R}(N)$ to be as small as possible with respect to $\mathbf{R}'(N)$, e.g., $\mathbf{R}(N) = \mathbf{R}'(N)$. To give a coarse asymptotic measure of security preserving properties, we classify reductions as either linear-preserving, poly-preserving, or weak-preserving.

Definition (security preserving reductions):

- *linear-preserving:* $\mathbf{R}(N) = N^{\mathcal{O}(1)} \cdot \mathcal{O}(\mathbf{R}'(N))$.

- *poly-preserving:* $\mathbf{R}(N) = N^{\mathcal{O}(1)} \cdot \mathbf{R}'(N)^{\mathcal{O}(1)}$.

- *weak-preserving:* $\mathbf{R}(N) = N^{\mathcal{O}(1)} \cdot \mathbf{R}'(N^{\mathcal{O}(1)})^{\mathcal{O}(1)}$. ♣

A linear-preserving reduction is more desirable than a poly-preserving reduction, which in turn is more desirable than a weak-preserving reduction. The difference between these types of guarantees increases dramatically as $\mathbf{R}'(N)$ grows. For example, if $\mathbf{R}'(N) = N^{\mathcal{O}(1)}$, then all types guarantee that $\mathbf{R}(N) = N^{\mathcal{O}(1)}$. In general though, for a weak-preserving reduction, $\mathbf{R}(N)$ may be much larger than $\mathbf{R}'(N)$ raised to any constant power. Thus, although a weak-preserving reduction does transfer some of the security of f to g, the transfer is extremely weak.

To exemplify the dramatic differences between the strengths of the reductions, let f be a one-way function and suppose we have a reduction from f to a pseudorandom generator g. Let A be an adversary for g with time-success ratio $\mathbf{R}'(N) = 2^{N^{1/2}}$. The following cases demonstrate how the strength of the reduction affects the time-success ratio $\mathbf{R}(N)$ of S^A for f.

- *linear-preserving:* Suppose $\mathbf{R}(N) = N^2 \cdot \mathbf{R}'(N)$. The time-success ratio of S^A for f is $\mathbf{R}(N) = N^2 \cdot 2^{N^{1/2}}$, which is not much larger than $\mathbf{R}'(N)$. Said differently, the reduction guarantees that if f is $(N^2 \cdot 2^{N^{1/2}})$-secure then g is $2^{N^{1/2}}$-secure.

- *poly-preserving:* Suppose $\mathbf{R}(N) = \mathbf{R}'(N)^2$. The time-success ratio of S^A for f is $\mathbf{R}(N) = 2^{2N^{1/2}}$, which is much larger than $\mathbf{R}'(N)$, but still only the square of $\mathbf{R}'(N)$. Said differently, the reduction guarantees that if f is $2^{2N^{1/2}}$-secure then g is $2^{N^{1/2}}$-secure.

- *weak-preserving:* Suppose $\mathbf{R}(N) = \mathbf{R}'(N^2)^2$. The time-success ratio of S^A for f is $\mathbf{R}(N) = 2^{2N}$, which is more than $\mathbf{R}'(N)$ raised to any constant power.

In the last example of a weak-preserving reduction, it is easy to see that there is an adversary for f with the stated time-success ratio. The adversary simply enumerates all possible private inputs to f of length N until it succeeds in finding an inverse. The running time of this adversary is 2^N multiplied by the time to evaluate f, and the adversary always succeeds. Thus, it is clear there is no 2^{2N}-secure one-way function f, and consequently this reduction is not strong enough to show that g is $2^{N^{1/2}}$-secure, no matter how secure f is assumed to be. The reduction does preserve some amount of security. For example, if f is $2^{2N^{1/2}}$-secure then g is $2^{N^{1/4}}$-secure. As can be seen by this example, the loss in security for this weak form of reduction is in general severe.

It turns out that the primary quantity that determines the strength of the reduction is the ratio $\mathbf{s}'(n)/\mathbf{s}(n)$. The bigger this ratio the more

Lecture 3

Overview

We define a weak one-way function and describe a weak-preserving reduction from a weak one-way function to a one-way function. We then describe several increasingly intricate linear-preserving reductions from a weak one-way permutation to a one-way permutation, where each subsequent reduction uses fewer public random bits than the previous.

Definition of a weak one-way function

Intuitively, f is a weak one-way function if it is hard to find an inverse of $f(x)$ for some significant but perhaps not very large fraction $x \in \{0,1\}^n$. (In contrast, for a one-way function it is hard to find an inverse of $f(x)$ for all but an insignificant fraction of the $x \in \{0,1\}^n$.)

Definition (weak one-way function): Let $f : \{0,1\}^n \to \{0,1\}^{\ell(n)}$ be a **P**-time function ensemble with security parameter $\|x\| = n$. The time bound and success probability of an adversary A for f are defined exactly the same way as for a one-way function. Let $\mathbf{w}(n)$ be a non-negligible parameter. Then, f is a $\mathbf{S}(n)$-*secure* $\mathbf{w}(n)$-*weak one-way function* if, for any adversary A with run time $T(n)$ and success probability $1 - \epsilon(n)$, either $T(n) \geq \mathbf{S}(n)$ or $\epsilon(n) \geq \mathbf{w}(n)$. ♣

Example : Define $f(x,y) = xy$, where $x,y \in \{2, \ldots, 2^n - 1\}$. The problem of inverting $f(x,y)$ consists of finding $x', y' \in \{2, \ldots, 2^n - 1\}$ such that $x'y' = xy$. Let $X, Y \in_{\mathcal{U}} \{2, \ldots, 2^n - 1\}$ be independent random variables. On average, $f(X,Y)$ is easy to invert, e.g., XY is an even number with probability $3/4$, in which case setting $x' = 2$ and $y' = XY/2$ inverts $f(X,Y)$. However, with probability approximately $1/n^2$ both X and Y are prime n-bit numbers. If there is no adversary that can factor the product of a pair of random n-bit prime numbers in time $T(n)$ then f is a $T(n)$-secure $(1/n^2)$-weak one-way function.

Strengthening weak functions

We now describe a weak-preserving reduction from a weak one-way function to a one-way function.

Uniform versus Non-uniform Reductions

In all of the above discussion, we only considered uniform reductions, i.e., S is an oracle adversary that can be computed by a **TM**. A non-uniform reduction is when S is an oracle circuit family, i.e., for each n, S_n is a circuit with oracle queries, and S_n^A is the same circuit where the oracle queries are computed by adversary A. In general, a uniform reduction is much more desirable then a non-uniform reduction, even if A is a circuit family.

To see the advantages, suppose we have a reduction from a one-way function f to a pseudorandom generator g, where the security of f is based on the difficulty of factoring the product of a pair of randomly chosen n-bit primes, and g is a pseudorandom generator constructed from f. Suppose that someone managed to find a **P**-time function ensemble A for distinguishing the output of g from truly random bits. If the reduction is uniform then S can be computed by a **TM** and thus S^A is a **P**-time function ensemble that factors a non-negligible fraction of products of n-bit primes. On the other hand, if the reduction is non-uniform then S is a circuit family and thus it may take time exponential in n to find the $n^{\mathcal{O}(1)}$ size circuit S_n such that S_n^A is a $n^{\mathcal{O}(1)}$ size circuit for factoring a non-negligible fraction of products of n-bit primes.

Even when the adversary A is a circuit family, a uniform reduction is still better than a non-uniform reduction. For example, suppose someone managed to find in time exponential in n a $n^{\mathcal{O}(1)}$ size circuit A_n that distinguishes the output of g from truly random bits. If S can be computed by a **TM** then, given A_n, S^{A_n} factors a non-negligible fraction of the products of n-bit primes in $n^{\mathcal{O}(1)}$ time. On the other hand, if S is an oracle circuit family then, even given the description of A_n, it still may take exponential time to construct the $n^{\mathcal{O}(1)}$ size oracle circuit S_n such that $S_n^{A_n}$ factors a non-negligible fraction of the products of n-bit primes in $n^{\mathcal{O}(1)}$ time.

The onus is on the adversary for a uniform reduction; the adversary may spend exponential time finding a $n^{\mathcal{O}(1)}$ size circuit A_n to break g if it exists. If the adversary is unsuccessful then we can use g securely, and on the other hand if the adversary finds A_n then in $n^{\mathcal{O}(1)}$ time (as opposed to possibly exponential time if the reduction is non-uniform) construct the $n^{\mathcal{O}(1)}$ size circuit S^{A_n} to solve the factoring problem.

Since the majority of the reductions we describe are uniform, we explicitly mention whether the reduction is uniform or non-uniform only when it is non-uniform.

Lecture 3

Overview

We define a weak one-way function and describe a weak-preserving reduction from a weak one-way function to a one-way function. We then describe several increasingly intricate linear-preserving reductions from a weak one-way permutation to a one-way permutation, where each subsequent reduction uses fewer public random bits than the previous.

Definition of a weak one-way function

Intuitively, f is a weak one-way function if it is hard to find an inverse of $f(x)$ for some significant but perhaps not very large fraction $x \in \{0,1\}^n$. (In contrast, for a one-way function it is hard to find an inverse of $f(x)$ for all but an insignificant fraction of the $x \in \{0,1\}^n$.)

Definition (weak one-way function): Let $f : \{0,1\}^n \to \{0,1\}^{\ell(n)}$ be a **P**-time function ensemble with security parameter $\|x\| = n$. The time bound and success probability of an adversary A for f are defined exactly the same way as for a one-way function. Let $\mathbf{w}(n)$ be a non-negligible parameter. Then, f is a $\mathbf{S}(n)$-*secure* $\mathbf{w}(n)$-*weak one-way function* if, for any adversary A with run time $T(n)$ and success probability $1 - \epsilon(n)$, either $T(n) \geq \mathbf{S}(n)$ or $\epsilon(n) \geq \mathbf{w}(n)$. ♣

Example : Define $f(x,y) = xy$, where $x, y \in \{2, \ldots, 2^n - 1\}$. The problem of inverting $f(x,y)$ consists of finding $x', y' \in \{2, \ldots, 2^n - 1\}$ such that $x'y' = xy$. Let $X, Y \in_{\mathcal{U}} \{2, \ldots, 2^n - 1\}$ be independent random variables. On average, $f(X,Y)$ is easy to invert, e.g., XY is an even number with probability $3/4$, in which case setting $x' = 2$ and $y' = XY/2$ inverts $f(X,Y)$. However, with probability approximately $1/n^2$ both X and Y are prime n-bit numbers. If there is no adversary that can factor the product of a pair of random n-bit prime numbers in time $T(n)$ then f is a $T(n)$-secure $(1/n^2)$-weak one-way function.

Strengthening weak functions

We now describe a weak-preserving reduction from a weak one-way function to a one-way function.

Construction 1 : Let $f : \{0,1\}^n \to \{0,1\}^{\ell(n)}$ be a $\mathbf{w}(n)$-weak one-way function. Let $N = 2n/\mathbf{w}(n)$, let $y \in \{0,1\}^{N \times n}$ and define the one-way function

$$g(y) = \langle f(y_1), \dots, f(y_N) \rangle$$

with security parameter $\| y \| = nN$.

Theorem 3.1 : If f is a weak one-way function then g is a one-way function. The reduction is weak-preserving.

———————∞———————

This theorem is proved below. We now describe a linear-preserving reduction from a weak one-way permutation to a one-way permutation. The basic structure of this reduction is similar to Construction 1. The main difference is that instead of the input to g being N private strings of length n each, the input to g is a single private string $x \in \{0,1\}^n$ and N public strings of length n each. The N public strings are used to generate N inputs of length n to f "sequentially".

Construction 2 : Let $f : \{0,1\}^n \to \{0,1\}^n$ be a $\mathbf{w}(n)$-weak one-way permutation. Let $N = 2n/\mathbf{w}(n)$, let $\pi = \langle \pi_1, \dots, \pi_N \rangle \in \{0,1\}^{N \times n}$ and define the one-way permutation

$$g(x, \pi) = \langle y_{N+1}, \pi \rangle$$

where $y_1 = x$ and, for all $i \in \{2, \dots, N+1\}$, $y_i = \pi_{i-1} \oplus f(y_{i-1})$. In this definition, π is a public input to g.

Theorem 3.2 : If f is a weak one-way permutation then g is a one-way permutation. The reduction is linear-preserving.

———————∞———————

The proof of Theorem 3.1 and Theorem 3.2 share the same basic structure, and are based on the development that follows. This development is described with respect to Construction 1, but the development is analogous for Construction 2. The difference in the strength of the reductions comes from the fact that in Construction 1 the length of the private input to g is y, which is much longer than n, whereas in Construction 2 the length of the private input to g is x, which is of length n.

Definition (H): Let $H = (\mathcal{F}, \mathcal{G}, E)$ be a bipartite graph, where $\mathcal{F} = \{0,1\}^n$ is the set of all inputs to f and $\mathcal{G} = \{0,1\}^{N \times n}$ is the set of all inputs to g. For all $x \in \mathcal{F}$ and for all $y \in \mathcal{G}$, $(x, y) \in E$ if there is some $i \in \{1, \dots, N\}$ such that $x = y_i$. (There are multiple edges in the case when $x = y_i = y_j$ for some $i \neq j$.) For each $x \in \mathcal{F}$, define the adjacency (multi)set of x as $\mathrm{adj}(x) = \{y \in \mathcal{G} : (x, y) \in E\}$ (y

appears in this multiset once for each edge $(x, y) \in E$). Similarly, define $\text{adj}(y) = \{x \in \mathcal{F} : (x, y) \in E\}$. Let $M = N2^{n(N-1)}$. Then, H is a (M, N)-regular graph, i.e., for each $x \in \mathcal{F}$, $\sharp\text{adj}(x) = M$ and for each $y \in \mathcal{G}$, $\sharp\text{adj}(y) = N$. ♣

The intuition behind the transfer of security from f to g is the following. Let A' be an adversary for inverting f and let

$$F = \{x \in \mathcal{F} : f(A'(f(x))) \neq f(x)\},$$

i.e., A' doesn't invert $f(x)$ for any $x \in F$. Let $X \in_{\mathcal{U}} \mathcal{F}$ and $Y \in_{\mathcal{U}} \mathcal{G}$. Let $\epsilon(n) = \Pr_X[X \in F]$ be the probability A' fails to invert f. Let A be the adversary that tries to invert $g(y)$ by using A' to try to invert $f(y_i)$, for each $i \in \{1, \ldots, N\}$. A successfully inverts $g(y)$ only if A' successfully inverts $f(y_i)$ for all $i \in \{1, \ldots, N\}$. We use the next definition to characterize the success probability of A.

Definition (forward expansion of H): H has (ϵ, δ)-*forward expansion* if the following is true: For all $F \subseteq \mathcal{F}$, if $\Pr_X[X \in F] \geq \epsilon$ then $\Pr_Y[\exists x \in F : x \in \text{adj}(Y)] \geq 1 - \delta$. ♣

From this definition and the above discussion, if H has $(\epsilon(n), \delta(n))$-forward expansion then the success probability of A is at most $\delta(n)$. In particular, H has $(\epsilon(n), \delta(n))$-forward expansion with $\delta(n) = (1 - \epsilon(n))^N$. Furthermore, if $\epsilon(n) \geq \mathbf{w}(n)/2$ then A succeeds in inverting with probability

$$\delta(n) \leq (1 - \mathbf{w}(n)/2)^N \leq e^{-N\mathbf{w}(n)/2} = e^{-n},$$

i.e., g is very secure *with respect to this particular adversary A*.

This particular adversary A shows that N cannot be much smaller, e.g. if $\epsilon(n) = \mathbf{w}(n)$ and if $N = 1/\mathbf{w}(n)$ then $\delta(n) \geq 1/e$. On the other hand, because the construction must be computable in $n^{\mathcal{O}(1)}$ time, $N = n^{\mathcal{O}(1)}$. These two constraints explain why we required $\mathbf{w}(n)$ to be a non-negligible parameter.

The reasoning above which shows that a particular adversary doesn't invert g must be generalized to show that there is *no* adversary that can invert g. We describe below an oracle adversary S with the property that if A is any breaking adversary for g then S^A is a breaking adversary for f. Let

$$G = \{y \in \mathcal{G} : g(A(g(y))) = g(y)\}$$

be the set A is able to invert and let $\delta(n) = \Pr_Y[Y \in G]$ be the success probability of A. For all $x \in \mathcal{F}$, let $Y(x) \in_{\mathcal{U}} \text{adj}(x)$. The following definition and theorem is the basis for the design of S.

Definition (reverse expansion of H): H has $(\epsilon, \delta, \gamma)$-*reverse expansion* if the following is true: For all $G \subseteq \mathcal{G}$, if $\Pr_Y[Y \in G] \geq \delta + \gamma$ then there is a set $F \subseteq \mathcal{F}$ such that:

- $\Pr_X[X \in F] \geq 1 - \epsilon$.

- For all $x \in F$, $\Pr_{Y(x)}[Y(x) \in G] \geq \gamma/N$.   ♣

The following provides the needed technical link between forward and reverse expansion.

Forward to Reverse Theorem :   For all $\epsilon, \delta > 0$, if H has (ϵ, δ)-forward expansion then, for all $\gamma > 0$, H has $(\epsilon, \delta, \gamma)$-reverse expansion.

PROOF: Fix $G \subseteq \mathcal{G}$ such that $\Pr_Y[Y \in G] \geq \delta + \gamma$. Define

$$F' = \{x \in \mathcal{F} : \Pr_{Y(x)}[Y(x) \in G] < \gamma/N\}$$

and

$$G' = \{y \in \mathcal{G} : \exists x \in F', y \in \mathrm{adj}(x)\}.$$

Suppose, for contradiction, that $\Pr_X[X \in F'] \geq \epsilon$. Then, since H has (ϵ, δ)-forward expansion, this implies that $\Pr_Y[Y \in G'] \geq 1 - \delta$. Let $G'' = G \cap G'$. From this it follows that $\Pr_Y[Y \in G''] \geq \gamma$ and thus

$$\sharp G'' \geq \gamma \cdot \sharp \mathcal{G} \geq \gamma \cdot \sharp G'.$$

Since all edges out of F' go into G', the number of edges out of F' is at most $N \cdot \sharp G'$. On the other hand, because $G'' \subseteq G'$ and because, for each $y \in G'$, $\mathrm{adj}(y) \cap F' \neq \emptyset$, the number of edges out of F' that go to G'' is at least $\sharp G'' \geq \gamma \cdot \sharp G'$. Let $X' \in_{\mathcal{U}} F'$. Thus,

$$\Pr_{X',Y(X')}[Y(X') \in G''] \geq \gamma/N.$$

This implies that there is an $x \in F'$ such that

$$\Pr_{Y(x)}[Y(x) \in G''] \geq \gamma/N.$$

Because $G'' \subseteq G$, this contradicts the definition of F'.   ■

PROOF of Theorem 3.1:

Suppose A is an adversary with time bound $T(n)$ and success probability $\delta(n)$ for g, and thus the time-success ratio is $\mathbf{R}(nN) = T(n)/\delta(n)$. Without much loss of generality we assume $\delta(n)/2 \geq e^{-n}$. Let $X \in_{\mathcal{U}} \mathcal{F}$ and $Y \in_{\mathcal{U}} \mathcal{G}$. Let $G \subseteq \mathcal{G}$ be the set of $y \in \mathcal{G}$ on which A is able to invert, and thus $\Pr_Y[Y \in G] = \delta(n)$. We describe the oracle adversary S and show that S^A inverts f with probability at least $1 - \mathbf{w}(n)$. The input to S^A is $f(x)$ where $x \in_{\mathcal{U}} \{0,1\}^n$.

Adversary S^A on input $f(x)$: .

Repeat $2nN/\delta(n)$ times

    Randomly choose $i \in_{\mathcal{U}} \{1, \ldots, N\}$.

    Randomly choose $y_1, \ldots, y_{i-1}, y_{i+1}, \ldots, y_N \in_{\mathcal{U}} \{0,1\}^n$.

    Set $\alpha = \langle f(y_1), \ldots, f(y_{i-1}), f(x), f(y_{i+1}), \ldots, f(y_N)\rangle$.

    $z \leftarrow A(\alpha)$

    If $f(z_i) = f(x)$ then output z_i.

Each execution of the repeat loop of S^A can be viewed in terms of H as choosing $\alpha = g(y)$ where $y \in_{\mathcal{U}} \operatorname{adj}(x)$. If $y \in G$, i.e., if A is able to invert $\alpha = g(y)$, then an inverse of $f(x)$ is successfully produced. H has $(\mathbf{w}(n)/2, \delta(n)/2)$-forward expansion by the choice of N and because of the assumption $\delta(n)/2 \geq e^{-n}$. It follows that H has $(\mathbf{w}(n)/2, \delta(n)/2, \delta(n)/2)$-reverse expansion from the Forward to Reverse Theorem. For each $x \in \mathcal{F}$ let $Y(x) \in_{\mathcal{U}} \operatorname{adj}(x)$. Because $\Pr_Y[Y \in G] = \delta(n)/2 + \delta(n)/2$, this implies there is a set $F \subseteq \mathcal{F}$ such that:

- $\Pr_X[X \in F] \geq 1 - \mathbf{w}(n)/2$.

- For each $x \in F$, $\Pr_{Y(x)}[Y(x) \in G] \geq \delta(n)/(2N)$.

Since the repeat loop is executed independently $2nN/\delta(n)$ times, S^A fails to find an inverse of $f(x)$ when $x \in F$ with probability at most

$$(1 - \delta(n)/(2N))^{2nN/\delta(n)} \leq e^{-n}.$$

The overall probability that S^A fails to find an inverse of $f(x)$ for randomly chosen $x \in_{\mathcal{U}} \mathcal{F}$ is at most

$$\Pr_X[X \notin F] + \Pr_X[X \in F] \cdot e^{-n} \leq \mathbf{w}(n)/2 + e^{-n} \leq \mathbf{w}(n).$$

The running time of S^A is dominated by the time for adversary A to compute the answers to the oracle queries. Each query to A takes $T(n)$

time, and thus the overall running time of S^A is

$$\mathcal{O}(nNT(n)/\delta(n)).$$

■

Theorem 3.1 is still true with minor modifications when the random input to f is part private and part public. Suppose the length of the public random input to f is $p(n)$. Then, g uses $p(n)N$ public random bits; these are partitioned into N strings of length $p(n)$ each, and these strings are used as the N public random strings needed for the N applications of f.

Generic Properties

For each construction described in this lecture, there is an associated (M, N)-regular bipartite graph $H = (\mathcal{F}, \mathcal{G}, E)$, where $\mathcal{F}$ is the set of inputs to f and $\mathcal{G}$ is the set of inputs to g, with the following properties:

(1) For any $x \in \mathcal{F}$, given $f(x)$ it is easy to produce $g(y)$, where $y \in_{\mathcal{U}}$ adj(x).

(2) For any $y \in$ adj(x), given any inverse z of $g(y)$, it is easy to compute an x' such that $f(x') = f(x)$.

Proof of Theorem 3.2

PROOF of Theorem 3.2: The proof is similar in spirit to the proof of Theorem 3.1. Suppose that A is an adversary with time bound $T(n)$ and success probability $\delta(n)$ for g, and thus the time-success ratio is $T(n)/\delta(n)$. Without much loss of generality we assume $\delta(n)/2 \geq e^{-n}$. On input $g(x, \pi)$, A finds x with probability $\delta(n)$ when $x \in_{\mathcal{U}} \{0, 1\}^n$ and $\pi \in_{\mathcal{U}} \{0, 1\}^{N \times n}$. We describe the oracle adversary S and show that S^A inverts f with probability at least $1 - \mathbf{w}(n)$. The input to S^A is $f(x)$ where $x \in_{\mathcal{U}} \{0, 1\}^n$.

Adversary S^A on input $f(x)$: .

Repeat $2nN/\delta(n)$ times

Randomly choose $i \in_{\mathcal{U}} \{2, \ldots, N + 1\}$.

Randomly choose $\pi \in_{\mathcal{U}} \{0,1\}^{N \times n}$

Let $y_i = \pi_{i-1} \oplus f(x)$.

Compute $y_{i+1} = f(y_i) \oplus \pi_i, \ldots, y_{N+1} = f(y_N) \oplus \pi_N$.

Compute $v_0 = A(y_{N+1}, \pi)$.

Compute $v_1 = \pi_0 \oplus f(v_0), \ldots, v_{i-1} = \pi_{i-2} \oplus f(v_{i-2})$.

if $f(v_{i-1}) = f(x)$ then output v_{i-1}.

We describe the edges E in the graph $H = (\mathcal{F}, \mathcal{G}, E)$ and verify that H has the required generic properties. Because f is a permutation, there is a unique sequence $\langle y_1(x, \pi), \ldots, y_{N+1}(x, \pi) \rangle$ associated with each $\langle x, \pi \rangle \in \mathcal{G}$, where $y_1(x, \pi) = x$ and $y_i(x, \pi) = \pi_{i-1} \oplus f(y_{i-1}(x, \pi))$ for all $i \in \{2, \ldots, N+1\}$. For each $i \in \{1, \ldots, N\}$ there is an edge in E between $y_i(x, \pi) \in \mathcal{F}$ and $\langle x, \pi \rangle \in \mathcal{G}$. Each input to g has degree N in the graph, and it is not hard to verify that each input to f has degree $N 2^{nN}$.

On input $f(x)$, S generates random $\pi \in_{\mathcal{U}} \{0,1\}^{N \times n}$, random $i \in_{\mathcal{U}} \{2, \ldots, N+1\}$ and produces $\langle y_{N+1}, \pi \rangle$ based on this. There is a unique input $\langle y_1, \pi \rangle$ to g such that $\langle g(y_1, \pi), \pi \rangle = \langle y_{N+1}, \pi \rangle$, and $\langle y_1, \pi \rangle$ defines the sequence $\langle y_1, \ldots, y_{N+1} \rangle$ of inputs to f, where $y_{i-1} = x$. Since π is uniformly distributed, $\langle y_1, \ldots, y_{N+1} \rangle$ is uniformly distributed conditional on $y_{i-1} = x$. The first generic property holds because $i \in_{\mathcal{U}} \{2, \ldots, N+1\}$. The second generic property holds because if A produces y_1 on input $\langle y_{N+1}, \pi \rangle$ then S^A produces x.

Let $X \in_{\mathcal{U}} \mathcal{F}$ and $Y \in_{\mathcal{U}} \mathcal{G}$. Let $F \subseteq \mathcal{F}$ and let $\epsilon = \Pr_X[X \in F]$. Then, $\Pr_Y[\exists x \in F : x \in \mathrm{adj}(Y)] = 1 - \delta$, where $\delta = (1 - \epsilon)^N$. Consequently, for any $\epsilon > 0$, H has $(\epsilon, (1-\epsilon)^N)$-forward expansion. The rest of the proof is exactly the same as the corresponding portion of the proof of Theorem 3.1 that appeals to the Forward to Reverse Theorem (with exactly the same setting of the parameters).

Finally, note that the running time of S^A is dominated by the time for adversary A to answer the oracle queries, and thus the overall running time is

$$\mathcal{O}(nNT(n)/\delta(n)).$$

■

Theorem 3.2 holds with minor modifications if the input to f is part private and part public. Suppose the length of the public input used by

f is $p(n)$. Then define

$$g(x, \pi, \rho) = \langle z_{N+1}, \pi, \rho \rangle,$$

where $x \in \{0,1\}^n$, $\pi \in \{0,1\}^{N \times n}$, $\rho \in \{0,1\}^{N \times p(n)}$, $z_1 = x$ and, for $i = 2, \ldots, N+1$, $z_i = \pi_{i-1} \oplus f(z_{i-1}, \rho_{i-1})$.

A linear-preserving reduction using less randomness

Although the reduction described in the preceding section is linear-pre-serving, it does have the drawback that the length of the public string used is large, and even worse this length depends linearly on $1/\mathbf{w}(n)$ for a $\mathbf{w}(n)$-weak one-way function. In this section, we describe a linear-preserving reduction that uses a much shorter public string.

In Construction 2, N public strings of length n each are used to construct N totally independent random inputs to f. The difference is that we use many fewer public random strings of length n in a recursive way to produce N random inputs to f that are somewhat independent of each other.

The reduction proceeds in two steps. In the first step we describe a linear-preserving reduction from a $\mathbf{w}(n)$-weak one-way permutation to a $(1/8)$-weak one-way permutation g. The second step reduces g to a one-way permutation h using the reduction described in Theorem 3.2.

Construction 3 : Let $f : \{0,1\}^n \to \{0,1\}^n$ be a $\mathbf{w}(n)$-weak one-way permutation. Let

$$\ell = \left\lceil 6 \log_{5/4}(1/\mathbf{w}(n)) \right\rceil$$

and let $N = 2^\ell$. Let $\pi = \langle \pi_1, \ldots, \pi_\ell \rangle \in \{0,1\}^{\ell \times n}$. Define

$$g(x, \pi_1) = \langle f(\pi_1 \oplus f(x)), \pi_1 \rangle.$$

For all $i = 2, \ldots, \ell$, recursively define

$$g(x, \pi_{\{1,\ldots,i\}}) = \langle g(\pi_i \oplus g(x, \pi_{\{1,\ldots,i-1\}}), \pi_{\{1,\ldots,i-1\}}), \pi_i \rangle.$$

The final output is

$$g(x, \pi) = g(x, \pi_{\{1,\ldots,\ell\}}).$$

--------∞--------

An iterative way to view this construction is the following. For all $i \in \{0,1\}^\ell \setminus \{0^\ell\}$ let

$$m_i = \min\{j \in \{1, \ldots, \ell\} : i_j = 1\}$$

be the index of the least significant bit of i that is equal to 1. Let $y_1 = x$. For all $i \in \{0,1\}^{\ell} \setminus \{0^{\ell}\}$ let $y_{i+1} = f(y_{i-1}) \oplus \pi_{m_i}$. Then, $g(x, \pi) = \langle f(y_N), \pi \rangle$.

This construction is similar to the second reduction, with some differences. It is still the case that, because f is a permutation, an input $\langle x, \pi \rangle$ to g uniquely determines a sequence of inputs $\langle y_1, \ldots, y_N \rangle$ to f. Although $\langle y_1, \ldots, y_N \rangle$ are not independent of each other as before, they are somewhat independent of each other.

Theorem 3.3 : If f is a weak one-way permutation then g is a $(1/8)$-weak one-way permutation. The reduction is linear-preserving.

PROOF: We first describe the graph H. An input $\langle x, \pi \rangle$ to g defines a sequence

$$y(x, \pi) = \langle y_1(x, \pi), \ldots, y_N(x, \pi) \rangle \in \{0,1\}^{N \times n},$$

where $y_1(x, \pi), \ldots, y_N(x, \pi)$ are the inputs to which the function f is applied when computing $g(x, \pi)$. There is an edge in H between $\langle x, \pi \rangle$ and $y_i(x, \pi)$ for all $i \in \{1, \ldots, N\}$. Given the inverse $\langle x, \pi \rangle$ of $g(x, \pi)$ it is easy to compute $y(x, \pi)$. Given $f(x)$ it is easy to generate an output of g that corresponds to a uniformly distributed $y = \langle y_1, \ldots, y_N \rangle$ where $x = y_i$ and $i \in_{\mathcal{U}} \{1, \ldots, N\}$: simply choose i uniformly, choose π uniformly, and then fix $f(x)$ as the output of the ith query to f and compute the rest of the sequence forward. This verifies that H has the generic properties. We now prove that H has forward expansion.

Lemma : H has $(\mathbf{w}(n)/2, 3/4)$-forward expansion.

PROOF: Let $X \in_{\mathcal{U}} \{0,1\}^n$ and

$$\Pi = \langle \Pi_1, \ldots, \Pi_l \rangle \in_{\mathcal{U}} \{0,1\}^{\ell \times n}.$$

Fix $F \subseteq \{0,1\}^n$ such that

$$\epsilon = \Pr_X[X \in F] \geq \mathbf{w}(n)/2.$$

Define $F_{\pi_{\emptyset}} = F$. Inductively, for all $i \geq 1$, define $F_{\pi_{\{1,\ldots,i\}}}$ as the set of all $x \in \{0,1\}^n$ such that in the computation of $g(x, \pi_{\{1,\ldots,i\}})$ at least one of the inputs to f is in F. Then,

$$F_{\pi_{\{1,\ldots,i\}}} = F_{\pi_{\{1,\ldots,i-1\}}} \cup g^{-1}(\pi_i \oplus F_{\pi_{\{1,\ldots,i-1\}}}, \pi_{\{1,\ldots,i-1\}}),$$

where $\pi_i \oplus F_{\pi_{\{1,\ldots,i-1\}}}$ is defined as the set $T = \{\pi_i \oplus x : x \in F_{\pi_{\{1,\ldots,i-1\}}}\}$ and $g^{-1}(T, \pi_{\{1,\ldots,i-1\}})$ is defined as the set $\{g^{-1}(x, \pi_{\{1,\ldots,i-1\}}) : x \in T\}$. Let

$$\epsilon_{\pi_{\emptyset}} = \epsilon$$

and for $i \geq 1$ let

$$\epsilon_{\pi_{\{1,\ldots,i\}}} = \Pr_{X}[X \in F_{\pi_{\{1,\ldots,i\}}}].$$

It is not hard to see that

$$\mathrm{E}_{\Pi}[\epsilon_{\Pi}] = \Pr_{(X,\Pi)}[\exists j \in \{1,\ldots,N\} : y_j(X,\Pi) \in F].$$

The proof of the lemma follows if $\mathrm{E}_{\Pi}[\epsilon_{\Pi}] \geq 1/4$. By Markov's inequality,

$$\mathrm{E}_{\Pi}[\epsilon_{\Pi}] \geq 1/2 \cdot \Pr_{\Pi}[\epsilon_{\Pi} \geq 1/2],$$

and thus it is sufficient to prove that $\Pr_{\Pi}[\epsilon_{\Pi} \geq 1/2] \geq 1/2$.

Let

$$F_{\pi_{\{1,\ldots,i-1\}}}(\pi_i) = F_{\pi_{\{1,\ldots,i\}}}$$

and

$$\epsilon_{\pi_{\{1,\ldots,i-1\}}}(\pi_i) = \epsilon_{\pi_{\{1,\ldots,i\}}}$$

be thought of as a function of π_i. Given $\pi_{\{1,\ldots,i-1\}}$, we say that π_i is *good* if

$$\epsilon_{\pi_{\{1,\ldots,i-1\}}}(\pi_i) \geq 5/4 \cdot \epsilon_{\pi_{\{1,\ldots,i-1\}}}.$$

Claim : If $\epsilon_{\pi_{\{1,\ldots,i-1\}}} \leq 1/2$ then $\Pr_{\Pi_i}[\Pi_i$ is good $] \geq 1/3$.

PROOF: Let

$$F'_{\pi_{\{1,\ldots,i-1\}}}(\pi_i) = g^{-1}(\pi_i \oplus F_{\pi_{\{1,\ldots,i-1\}}}, \pi_{\{1,\ldots,i-1\}}).$$

Then,

$$F_{\pi_{\{1,\ldots,i-1\}}}(\pi_i) = F_{\pi_{\{1,\ldots,i-1\}}} \cup F'_{\pi_{\{1,\ldots,i-1\}}}(\pi_i).$$

Because g is a permutation it follows that, for all values of π_i,

$$\Pr_{X}[X \in F'_{\pi_{\{1,\ldots,i-1\}}}(\pi_i)] = \Pr_{X}[X \in F_{\pi_{\{1,\ldots,i\}}}] = \epsilon_{\pi_{\{1,\ldots,i-1\}}}.$$

Furthermore, for each fixed $x \in \{0,1\}^n$,

$$\Pr_{\Pi_i}[x \in F'_{\pi_{\{1,\ldots,i-1\}}}(\Pi_i)] = \epsilon_{\pi_{\{1,\ldots,i-1\}}}.$$

From this it follows that the event $X \in F'_{\pi_{\{1,\ldots,i-1\}}}(\Pi_i)$ is independent of the event $X \in F_{\pi_{\{1,\ldots,i-1\}}}$, and each of these events occurs with probability $\epsilon_{\pi_{\{1,\ldots,i-1\}}}$. Thus, the probability that at least one of these two events occurs is the sum of their probabilities minus the product of their probabilities, i.e.,

$$
\begin{aligned}
\mathrm{E}_{\Pi_i} & \left[\epsilon_{\pi_{\{1,\ldots,i-1\}}}(\Pi_i)\right] \\
&= \mathrm{Pr}_{X,\Pi_i}[X \in F_{\pi_{\{1,\ldots,i-1\}}} \text{ or } X \in F'_{\pi_{\{1,\ldots,i-1\}}}(\Pi_i)] \\
&= \epsilon_{\pi_{\{1,\ldots,i-1\}}}(2 - \epsilon_{\pi_{\{1,\ldots,i-1\}}}).
\end{aligned}
$$

If $\epsilon_{\pi_{\{1,\ldots,i-1\}}} \leq 1/2$ then this implies that

$$
\mathrm{E}_{\Pi_i}[\epsilon_{\pi_{\{1,\ldots,i-1\}}}(\Pi_i)] \geq 3/2 \cdot \epsilon_{\pi_{\{1,\ldots,i-1\}}}. \tag{1}
$$

Note that for any value of π_i,

$$
\epsilon_{\pi_{\{1,\ldots,i-1\}}}(\pi_i) \leq 2\epsilon_{\pi_{\{1,\ldots,i-1\}}}. \tag{2}
$$

Let

$$
p = \mathrm{Pr}_{\Pi_i}[\epsilon_{\pi_{\{1,\ldots,i-1\}}}(\Pi_i) \leq 5/4 \cdot \epsilon_{\pi_{\{1,\ldots,i-1\}}}].
$$

From the definition of p and from equation (2) it follows that

$$
\mathrm{E}_{\Pi_i}[\epsilon_{\pi_{\{1,\ldots,i-1\}}}(\Pi_i)] \leq (5/4 \cdot p + 2(1-p)) \cdot \epsilon_{\pi_{\{1,\ldots,i-1\}}}.
$$

From this and equation (1) it follows that $p \leq 2/3$. This complete the proof of the claim. ∎

We complete the proof of the lemma. For any value of π_i, $\epsilon_{\pi_{\{1,\ldots,i-1\}}}(\pi_i) \geq \epsilon_{\pi_{\{1,\ldots,i-1\}}}$. Thus, even when π_i is not good, the value of $\epsilon_{\pi_{\{1,\ldots,i-1\}}}(\pi_i)$ is at least as large as $\epsilon_{\pi_{\{1,\ldots,i-1\}}}$. Consider a coin that has probability $1/3$ at each toss of landing heads. From the claim it follows that

$$
\mathrm{Pr}_{\Pi_{\{1,\ldots,i\}}}[\epsilon_{\Pi_{\{1,\ldots,i\}}} \geq \min\{(5/4)^j \cdot \epsilon, 1/2\}]
$$

is lower bounded by the probability that there at least j outcomes of heads in i independent tosses of the coin. From this and noting that

$$
(5/4)^{\ell/6} \cdot \epsilon \geq 1/\mathbf{w}(n) \cdot \mathbf{w}(n)/2 \geq 1/2,
$$

it follows that

$$
\mathrm{Pr}_{\Pi}[\epsilon_{\Pi} \geq 1/2]
$$

is at least the probability that there at least $\ell/6$ outcomes of heads out of ℓ independent tosses of the coin. Note that the expected number of heads in ℓ coin tosses is $\ell/3$. It is easy to verify using Chernoff bounds (see page 12 in the Preliminaries) that the probability that at least $\ell/6$ of the ℓ coin tosses land heads is at least $1/2$. This completes the proof of the lemma. ∎

We complete the proof of the theorem. The lemma and the Forward to Reverse Theorem implies that H has $(\mathbf{w}(n)/2, 3/4, 1/8)$-reverse expansion. Suppose that A is an adversary for g with time bound $T(n)$ and success probability at least $7/8 = 3/4 + 1/8$ for g. The oracle adversary S for this reduction is practically the same as described in Theorem 3.2; the repeat loop is iterated $\mathcal{O}(nN)$ times, where each time a random output of g given $f(x)$ is chosen as described above. A detailed description of S is omitted. The run time of S^A is dominated by the time for adversary A to answer the oracle queries, and thus the overall running time is

$$\mathcal{O}(nNT(n)).$$

■

The final step in the reduction is to go from $(1/8)$-weak one-way permutation g to a one-way permutation h using Construction 2, where g uses a public random string of length $m(n) = \mathcal{O}(n\log(1/\mathbf{w}(n)))$. When using Construction 2 to go from g to h, we set $N = \log(1/R_g(n)) \leq n$ and partition the public random string into N blocks of length $n + m(n)$. Thus, the overall construction uses $\mathcal{O}(n^2 \log(1/\mathbf{w}(n)))$ public random bits, as opposed to $\mathcal{O}(n^2/\mathbf{w}(n))$ for Construction 2. It is not hard to verify that the overall reduction from f to h is linear-preserving.

A linear-preserving reduction using expander graphs

In this section, we describe a linear-preserving reduction from a weak one-way permutation f to a one-way permutation h that uses only a linear number of public random bits overall. We only sketch the similarities and differences between the reduction described here and previously described reductions, leaving out all details and proofs.

The construction builds on the ideas developed in Construction 3. As in that reduction, the reduction described here proceeds in two steps: The first step is a linear-preserving reduction from a $\mathbf{w}(n)$-weak one-way permutation f to a $(1/2)$-weak one-way permutation g; The second step reduces g to a one-way permutation h. As in Construction 3, the first step is recursive and uses $\mathcal{O}(\log(\mathbf{w}(n)))$ public random strings, but these are of constant length each instead of length n. The second step is iterative, but uses only $\mathcal{O}(n)$ public random bits. The number of public random bits used overall in the entire construction from f to h is linear in n, and thus the reduction is poly-preserving even when the public random string is considered as part of the private input to h.

The overall structure of the first step of the reduction uses the same recursive structure as used in Construction 3 (with roughly the same parameters, i.e., $\ell = \mathcal{O}(\log(1/\mathbf{w}(n))$ and $N = 2^\ell$). The main idea is to use a constant number of bits ρ_i to replace each n bit string π_i. This is done by defining a constant degree d expander graph on the set of all inputs $\{0, 1\}^n$ to f, and letting $\rho_i \in \{1, \ldots, d\}$ select one of the d possible edges out of a vertex. Thus, whenever we would have computed $z' = \pi \oplus z$ in Construction 3, in this reduction we follow the edge labeled ρ_i out of $z \in \{0, 1\}^n$ to a node $z' \in \{0, 1\}^n$. The analysis that shows that the construction graph has (ϵ, δ)-forward expansion when $\epsilon = \mathbf{w}(n)/2$ and $\delta = 3/4$ is similar to the analysis of Construction 3, and relies on properties of the expander graph. The main difference is how the claim used to prove Theorem 3.3 is shown. The property we use is that for any subset F of at most half the vertices, the neighborhood set of F (all those vertices adjacent to at least one vertex in F, including F itself) is of size at least $(1 + \theta) \cdot \sharp F$ (for some constant $\theta > 0$ that depends on the expansion properties of the graph.) The remaining details of this analysis are omitted.

The second step is a reduction from g to a one-way permutation h. If we use the same idea as used in the previous section for the second step, the total number of public random bits would be $\mathcal{O}(n \log(1/\mathbf{w}(n)))$, because we would use $\mathcal{O}(\log(1/\mathbf{w}(n)))$ public random bits in each of the n queries to the function g when computing h.

We can improve the construction of h from g to use only $\mathcal{O}(n)$ public random bits as follows. Define a constant degree d' expander graph H', where there is a vertex in H' for each $\langle x, \rho_1, \ldots, \rho_\ell \rangle$, where $x \in \{0, 1\}^n$ is the private input to g and $\rho_1, \ldots, \rho_\ell \in \{1, \ldots, d\}$ are a possible setting of the public random bits used by g. Let $m = cn$ for some constant $c > 1$ and let $\alpha_1, \ldots, \alpha_m \in_{\mathcal{U}} \{1, \ldots, d'\}$ be m randomly and independently chosen edge labels for the graph H'. Then, define

$$h(x, \rho_1, \ldots, \rho_\ell, \alpha_1, \ldots, \alpha_m)$$

as follows. $y_1 = x$ and $\psi_1 = \langle \rho_1, \ldots, \rho_\ell \rangle$. Define $\langle y_{i+1}, \psi_{i+1} \rangle$ inductively from $\langle y_i, \psi_i \rangle$ as follows: Compute $g(y_i, \psi_i)$, follow the edge labeled α_i from node $\langle g(y_i, \psi_i), \psi_i \rangle$ to node $\langle y_{i+1}, \psi_{i+1} \rangle$.

Let $F \subset \{0, 1\}^n$ and $\epsilon = \sharp F/2^n = \mathbf{w}(n)/2$. The overall proof directly shows that in the two step reduction from f to h the graph for the reduction has (ϵ, δ)-forward expansion with respect to this value of ϵ and δ exponentially small in n. The intuitive reason for this is that each application of g defines $N \approx 1/\mathbf{w}(n)$ inputs to f, and at least one of these inputs is in F with constant probability by the forward expansion of g. Thus, since when we compute h we are basically taking a random walk on the expander graph H' of length m, by properties of expander graphs it is only with probability exponentially small in n that none of these mN inputs to g are in F.

The reduction as described is valid only when f is a permutation. As is the case with the previously described reductions, this reduction can also be generalized to the case where f is a weak one-way regular function (See page 91 for a definition of a regular function ensemble.)

Research Problem 1 : The last three constructions are better than Construction 1 in terms of their security preserving properties, but they have the disadvantage that they only apply to permutations. All three constructions can be generalized to apply to regular functions. (See page 91 for a definition of a regular function ensemble.) However, there is no known linear-preserving (or even poly-preserving) reduction from a weak one-way function to a one-way function for functions with no restrictions, and the design of such a reduction is a good research problem. ♠

Research Problem 2 : Both Constructions 1 and 2 require applying f to many inputs, but the parallel time to compute Construction 1 is proportional to the time to compute f, whereas for Construction 2 the parallel time is proportional to N multiplied by the time to compute f. An interesting research problem is to design a linear-preserving (or poly-preserving) reduction from weak one-way permutation f to one-way permutation g such that the parallel time for computing g is comparable to that for computing f, e.g., proportional to the time to compute f or logarithmic in the input size multiplied by the time to compute f. (This is a harder problem than Research Problem 1 for general functions.) ♠

Lecture 4

Overview

Using the random self-reducibility property of the discrete log problem, we show that if it is easy to invert a significant fraction of its inputs then it is easy to invert all its inputs. The definition of a pseudorandom generator and a definition of pseudorandomness based on the next bit test are given, and the two definitions are shown to be equivalent. A construction of a pseudorandom generator that produces a long output from a pseudorandom generator that produces an output one bit longer than its input is given.

The discrete log problem and random self-reducibility

Partially justified by the example on page 26, we introduced the time-success ratio as the single measure of how well an adversary breaks a primitive. This example shows that one can decrease the average run time of the adversary with only a corresponding linear decrease in the success probability. In general, it is not known how to go the other direction, i.e., how to increase the run time of the adversary with a corresponding increase in the success probability. The following discussion shows that this is possible for the discrete log function previously introduced on page 17.

The discrete log function : Let p be a prime, $\|p\| = n$, and let g be a generator of $\mathbb{Z}_p^*$. Define $f(p, g, x) = \langle p, g, g^x \bmod p \rangle$, where $x \in \mathbb{Z}_{p-1}$. Both p and g are public inputs and x is a private input, and thus the security parameter is $\|x\| = n$.

Self-Reducible Theorem : There is an oracle adversary S with the following properties. Let $X \in_{\mathcal{U}} \mathbb{Z}_p^*$, and let $A : \{0,1\}^n \times \{0,1\}^n \times \{0,1\}^n \to \{0,1\}^n$ be an adversary for inverting the discrete log function f with success probability

$$\delta(n) = \Pr_X[A(f(p, g, X)) = f(p, g, X)],$$

and with run time $T(n)$. Then, S^A has success probability at least $1/2$ for inverting $f(p, g, X)$ and the run time of S^A is $\mathcal{O}(T(n)/\delta(n))$.

PROOF: S^A works as follows on input $\langle p, g, y \rangle$, where $y = g^x \bmod p$.

Adversary S^A on input $\langle p, g, y \rangle$:

Repeat $1/\delta(n)$ times

    Choose $x' \in_{\mathcal{U}} \mathcal{Z}_{p-1}$

    Set $y' = yg^{x'} \bmod p$.

    Set $x'' = A(y', p, g)$.

    If $g^{x''} = y' \bmod p$ then output $(x'' - x') \bmod (p - 1)$ and stop.

Output 0^n.

Let $X' \in_{\mathcal{U}} \mathcal{Z}_{p-1}$. Then $Y' \in_{\mathcal{U}} \mathcal{Z}_p^*$, where $Y' = yg^{X'} \bmod p$. Thus, on any input y, the probability that A fails to invert y' in all of the independent trials is at most $(1 - \delta(n))^{1/\delta(n)} \leq 1/2$. Thus, with probability at least $1/2$, in at least one of the trials A successfully finds an x'' such that $g^{x''} = y' \bmod p$. Because $g^{x''} = g^{x+x'} \bmod p$ and because in general $g^{i(p-1)+j} = g^j \bmod p$ for any pair of integers i and j, it follows that $x = (x'' - x') \bmod (p - 1)$. Thus, S^A succeeds in inverting any fixed input y with probability at least $1/2$. The time bound for S^A is

$$\mathcal{O}(T(n)/\delta(n)).$$

■

Random Self-Reducibility : The proof works because inverting any output of the discrete log function can be reduced to inverting a random output. This property is called random self-reducibility. In the above example, the original instance is reduced to one random instance of the same problem. For such problems, either an overwhelming fraction of the instances are hard, or else all instances are easy.

---

pseudorandom generators and next bit unpredictability

We now give the definitions of a pseudorandom generator and of a next bit unpredictable function.

Definition (pseudorandom generator): Let $g : \{0, 1\}^n \rightarrow \{0, 1\}^{\ell(n)}$ be a **P**-time function ensemble, where $\ell(n) > n$. The input to g is private, and thus the security parameter $\mathbf{S}(n)$ of g is n. The stretching parameter of g is $\ell(n) - n$. Let $X \in_{\mathcal{U}} \{0,1\}^n$ and $Z \in_{\mathcal{U}} \{0,1\}^{\ell(n)}$. The success probability (distinguishing probability) of adversary A for g is

$$\delta(n) = |\Pr_X[A(g(X)) = 1] - \Pr_Z[A(Z) = 1]|.$$

Then, g is a $\mathbf{S}(n)$-*secure pseudorandom generator* if every adversary has time-success ratio at least $\mathbf{S}(n)$. ♣

Definition (next bit unpredictable): Let $g : \{0,1\}^n \to \{0,1\}^{\ell(n)}$ be a $\mathbf{P}$-time function ensemble, where $\ell(n) > n$. The input to g is private, and thus the security parameter of g is n. Let $X \in_{\mathcal{U}} \{0,1\}^n$ and $I \in_{\mathcal{U}} \{1, \ldots, \ell(n)\}$. The success probability (prediction probability) of A for g is

$$\delta(n) = \Pr_{I,X}[A(I, g(X)_{\{1,\ldots,I-1\}}) = g(X)_I] - 1/2.$$

Then, g is a $\mathbf{S}(n)$-*secure next-bit unpredictable* if every adversary has time-success ratio at least $\mathbf{S}(n)$. ♣

In this definition, $\delta(n)$ measures how well A can predict the ith bit $g(X)_i$ of the output given the first $i - 1$ bits $g(X)_{\{1,\ldots,i-1\}}$, for random $i \in_{\mathcal{U}} \{1, \ldots, \ell(n)\}$.

Theorem 4.1 : Let $g : \{0,1\}^n \to \{0,1\}^{\ell(n)}$ be a $\mathbf{P}$-time function ensemble, where $\ell(n) > n$. Then, g is a pseudorandom generator if and only if g is next-bit unpredictable. The reduction is linear-preserving in both directions.

Exercise 10 : Prove the first part of Theorem 4.1. More precisely, prove there is an oracle adversary S such that if A is an adversary for g with time-success ratio $\mathbf{R}'(n)$ in terms of next-bit unpredictability then S^A is an adversary for g with time-success ratio $\mathbf{R}(n)$ in terms of pseudorandomness, where $\mathbf{R}(n) = n^{\mathcal{O}(1)} \cdot \mathcal{O}(\mathbf{R}'(n))$. ♠

Exercise 11 : Prove the second part of Theorem 4.1. More precisely, prove there is an oracle adversary S such that if A is an adversary for g with time-success ratio $\mathbf{R}'(n)$ in terms of pseudorandomness then S^A is an adversary for g with time-success ratio $\mathbf{R}(n)$ in terms of next-bit unpredictability, where $\mathbf{R}(n) = n^{\mathcal{O}(1)} \cdot \mathcal{O}(\mathbf{R}'(n))$.

Hint : Let $\delta(n)$ be the the success probability of A. Let $X \in_{\mathcal{U}} \{0,1\}^n$ and $Y \in_{\mathcal{U}} \{0,1\}^{\ell(n)}$. Consider the following sequence of distributions.

$$\mathcal{D}_0 = Y$$

$$\mathcal{D}_1 = \langle g(X)_1, Y_{\{2,\ldots,\ell(n)\}}\rangle$$

$\vdots$

$$\mathcal{D}_i = \langle g(X)_{\{1,\ldots,i\}}, Y_{\{i+1,\ldots,\ell(n)\}}\rangle$$

$\vdots$

$$\mathcal{D}_{\ell(n)} = g(X)$$

For each $i = 0, \ldots, \ell(n)$, let δ_i be the probability that A outputs 1 on an input randomly chosen according to $\mathcal{D}_i$. Let $I \in_{\mathcal{U}} \{1, \ldots, \ell(n)\}$. Because $\delta(n) = \delta_{\ell(n)} - \delta_0$, it follows that

$$\mathrm{E}_I[\delta_I - \delta_{I-1}] = \delta(n)/\ell(n).$$

This suggests a way to predict the next bit.   ♠

Stretching the output of a pseudorandom generator

One concern with using a pseudorandom generator is that the length of its output may not be long enough for the application in mind. A pseudorandom generator may stretch by only a single bit, whereas in many applications it is important that the length of its output be much longer than its input. For example, when a pseudorandom generator is used in a private key cryptosystem, the length its output should be at least as long as the total length of the messages to be encrypted. Furthermore, the total length of the messages may not be known beforehand, and thus it is important to be able to produce the bits of the pseudorandom generator in an on-line fashion. The following theorem shows that it is possible to construct a pseudorandom generator that stretches by an arbitrary polynomial amount in an on-line fashion from a pseudorandom generator that stretches by a single bit. We only prove the theorem for the case when the entire input is private, but the proof is no more difficult when the input consists of both a private and a public part.

Stretching construction :   Let $g : \{0,1\}^n \to \{0,1\}^{n+1}$ be a pseudorandom generator, i.e., g stretches the input by one bit. Define $g^0(x) = x$, $g^1(x) = g(x)$ and, for all $i \geq 1$,

$$g^{i+1}(x) = \langle g(x)_1, g^i(g(x)_{\{2,\ldots,n+1\}})\rangle.$$

Stretching Theorem :   Let $\ell(n) > n$ be a polynomial parameter. If g is a pseudorandom generator then $g^{\ell(n)}$ is a pseudorandom generator. The reduction is linear-preserving.

PROOF: Let $X \in_{\mathcal{U}} \{0,1\}^n$ and $Z \in_{\mathcal{U}} \{0,1\}^{n+\ell(n)}$. Let A be an adversary for $g^{\ell(n)}(X)$ with success probability

$$\delta(n) = \Pr[A(g^{\ell(n)}(X)) = 1] - \Pr[A(Z) = 1]$$

and time bound $T(n)$. We describe an oracle adversary S such that S^A has time bound $\mathcal{O}(T(n))$ and such that the success probability of S^A for g is $\delta(n)/\ell(n)$. Let $Y \in_{\mathcal{U}} \{0,1\}^{\ell(n)}$. Consider the following sequence of distributions:

$$\mathcal{D}_0 = \langle Y, X \rangle$$

$$\mathcal{D}_1 = \langle Y_{\{1,\ldots,\ell(n)-1\}}, g^1(X) \rangle$$

$$\vdots$$

$$\mathcal{D}_i = \langle Y_{\{1,\ldots,\ell(n)-i\}}, g^i(X) \rangle$$

$$\vdots$$

$$\mathcal{D}_{\ell(n)} = g^{\ell(n)}(X)$$

For each $i = 0, \ldots, \ell(n)$, let δ_i be the probability that A outputs 1 on an input randomly chosen according to $\mathcal{D}_i$. Let $I \in_{\mathcal{U}} \{1, \ldots, \ell(n)\}$ Because $\delta(n) = \delta_{\ell(n)} - \delta_0$, it follows that

$$\mathrm{E}_I[\delta_I - \delta_{I-1}] = \delta(n)/\ell(n).$$

We describe the oracle adversary S. The input to S^A is $u \in \{0,1\}^{n+1}$.

Adversary S^A on input u : .

    Randomly choose $i \in_{\mathcal{U}} \{1, \ldots, \ell(n)\}$.

    Randomly choose $y \in_{\mathcal{U}} \{0,1\}^{\ell(n)-i}$.

    Set $\alpha = \langle y, u_1, g^{i-1}(u_{\{2,\ldots,n+1\}}) \rangle$.

    Output $A(\alpha)$

For fixed $i \in \{1, \ldots, \ell(n)\}$, if $u \in_{\mathcal{U}} \{0,1\}^{n+1}$ then the probability S^A produces 1 is δ_{i-1}, whereas if $u \in_{g(X)} \{0,1\}^{n+1}$ then this probability is δ_i. Since $i \in_{\mathcal{U}} \{1, \ldots, \ell(n)\}$, it follows that the success probability (distinguishing probability) of S^A is $\delta(n)/\ell(n)$. ■

If g is a pseudorandom generator that stretches by more than one bit, e.g., by $p(n)$ bits, then the stretching construction generalizes in an obvious way to produce a pseudorandom generator $g^{\ell(n)}$ that stretches by $\ell(n)p(n)$.

Based on the Stretching Theorem, the following algorithm can be used to produce $\ell(n)$ pseudorandom bits from a pseudorandom generator g that stretches by one bit. In this implementation, the total amount of private memory used is the same as the length of the private input to g. Initially, $x \in_{\mathcal{U}} \{0,1\}^n$ is produced using the private random source and stored in the private memory.

Stretching Algorithm :

For $i = 1, \ldots, \ell(n)$ do

> Move x from the private memory to the private computational device.
>
> Compute $g(x)$ in the private computational device.
>
> Store $a_i = g(x)_1$ in the public memory.
>
> Replace x with $g(x)_{\{2,\ldots,n+1\}}$ in the private memory.

At the termination of the algorithm, the string $a = \langle a_1, \ldots, a_{\ell(n)} \rangle$ stored in the public memory unit is the pseudorandom string. The next bit in the pseudorandom sequence is computed based on the n bits currently stored in the private memory. During periods when there is no demand for the next bit, the private computational device need not be protected as long as its contents are destroyed after each use.

Exercise 12 : Let $T(n)$ be a set of functions and let **DTIME**$(T(n))$ be the class of all languages $L : \{0,1\}^n \to \{0,1\}$ such that there is a $t(n) \in T(n)$ and a **TM** $M : \{0,1\}^n \to \{0,1\}$ with running time bounded by $t(n)$ that determines membership in L, i.e., for all $x \in L$, $M(x) = 1$ and for all $x \notin L$, $M(x) = 0$. Given a function $\mathbf{S}(n)$, define $\mathcal{S}(n)$ as the class of all functions

$$\{s(n) : \mathbf{S}(\log(s(n))) = n^{\mathcal{O}(1)}\}.$$

Prove that if there is a $\mathbf{S}(n)$-secure pseudorandom generator then **BPP** $\subseteq$ **DTIME**$(\mathcal{S}(n))$. Thus, for example, if there is a $2^{\sqrt{n}}$-secure pseudorandom generator then **BPP** $\subseteq$ **DTIME**$(2^{\mathcal{O}(\log^2(n))})$. Note that you will have to assume that the pseudorandom generator is $S(n)$-secure against non-uniform adversaries to solve this exercise. (Do you see why?)   ♠

Research Problem 3 : Is there a poly-preserving or linear-preserving reduction from a one-stretching pseudorandom generator to an n-stretching pseudorandom generator that can be computed fast in parallel? ♠

Lecture 5

Overview

We introduce a paradigm for derandomizing probabilistic algorithms for a variety of problems. This approach is of central importance for many of the constructions introduced in subsequent lectures.

The Paradigm

The paradigm consists of two complementary parts. The first part is to design a probabilistic algorithm described by a sequence of random variables so that the analysis is valid assuming limited independence between the random variables. The second part is the design of a small probability space for the random variables such that they are somewhat independent of each other. Thus, the random variables used by the algorithm can be generated according to the small space and still the analysis of the algorithm holds.

Limited Independence Probability Spaces

We describe constructions of probability spaces that induce a pairwise independent distribution on a sequence of random variables. The advantages are that the size of the space is small and it can be constructed with properties that can be exploited by an algorithm. It turns out that the analysis of many probabilistic algorithms requires only pairwise independence between the random variables.

Definition (pairwise independence and k-wise independence):
Let $X_1, \ldots, X_m$ be a sequence of random variables with values in a set S. The random variables are *pairwise independent* if, for all $1 \leq i < j \leq m$ and for all $\alpha, \beta \in S$,

$$\Pr[X_i = \alpha \wedge X_j = \beta] = \Pr[X_i = \alpha] \cdot \Pr[X_j = \beta].$$

There is no requirement that larger subsets are independent, e.g., the variables are not independent in triples in the constructions below for pairwise independent random variables. More generally, we say they are *k-wise independent* if, for all $1 \leq i_1 < \cdots < i_k \leq m$ and for all

$\alpha_1, \ldots, \alpha_k \in S$,

$$\Pr[X_{i_1} = \alpha_1 \wedge \cdots \wedge X_{i_k} = \alpha_k] = \Pr[X_{i_1} = \alpha_1] \cdots \Pr[X_{i_k} = \alpha_k].$$

♣

Modulo Prime Space : Let p be a prime number. The sample space is the set of all pairs $S = \{\langle a, b \rangle : a, b \in \mathbb{Z}_p\}$, and the distribution on the sample points is uniform, i.e., $\langle A, B \rangle \in_{\mathcal{U}} S$. Let ζ be an indeterminate and consider the polynomial

$$p_{a,b}(\zeta) = (a\zeta + b) \bmod p,$$

where $\langle a, b \rangle \in S$. For all $i \in \mathbb{Z}_p$, define random variable

$$X_i(A, B) = p_{A,B}(i).$$

For brevity, we sometimes use X_i in place of $X_i(A, B)$.

Claim : $\langle X_0, \ldots, X_{p-1} \rangle$ are uniformly distributed in $\mathbb{Z}_p$ and pairwise independent.

PROOF: For any pair $i, j \in \mathbb{Z}_p$, $i \neq j$, and for any pair of values $\alpha, \beta \in \mathbb{Z}_p$, there is a unique solution $a, b \in \mathbb{Z}_p$ to the pair of equations:

- $p_{a,b}(i) = \alpha$.

- $p_{a,b}(j) = \beta$.

Thus, $\Pr_{A,B}[X_i(A, B) = \alpha \wedge X_j(A, B) = \beta] = 1/p^2$. ■

Exercise 13 : Let p be a prime number and let $m \leq p$. Generalize the Modulo Prime Space to a probability space where $X_0, \ldots, X_{m-1} \in_{\mathcal{U}} \mathbb{Z}_p$ are k-wise independent, where the size of the probability space is p^k. ♠

The Modulo Prime Space can be generalized as follows.

Linear Polynomial Space : Let $\mathcal{F}$ be any finite field and consider the polynomial

$$p_{a,b}(\zeta) = a\zeta + b$$

over $\mathcal{F}$, where $a, b \in \mathcal{F}$. Identify the integers $\{0, \ldots, \sharp\mathcal{F} - 1\}$ with the elements of $\mathcal{F}$. The sample space is $S = \{\langle a, b \rangle : a, b \in \mathcal{F}\}$ and the distribution on S is $\langle A, B \rangle \in_{\mathcal{U}} S$. For all $i \in \mathcal{F}$, define random variable

$$X_i(A, B) = p_{A,B}(i),$$

where i on the left side of the equality is treated as an index and on the right side of the equality it is the corresponding element of $\mathcal{F}$.

——————∞——————

The random variables $\langle X_0, \ldots, X_{\sharp\mathcal{F}-1} \rangle$ are uniformly distributed in $\mathcal{F}$ and pairwise independent. A field with nice properties is $\mathrm{GF}[2^n]$, the Galois field with 2^n elements.

Mapping between $\{0, 1\}^n$ and $\mathrm{GF}[2^n]$: There is a natural mapping between $\{0, 1\}^n$ and polynomials in one variable ζ of degree $n - 1$ over $\mathrm{GF}[2]$. Namely, if $a \in \{0, 1\}^n$ and $\langle a_0, \ldots, a_{n-1} \rangle$ are the bits of a then the corresponding polynomial is

$$a(\zeta) = \sum_{i=0}^{n-1} a_i \zeta^i.$$

Each element of $\mathrm{GF}[2^n]$ can be represented by an n-bit string. Let $a \in \{0, 1\}^n$ and $b \in \{0, 1\}^n$ and let $a(\zeta)$ and $b(\zeta)$ be the corresponding polynomials. Computing $a + b$ over $\mathrm{GF}[2^n]$ consists of computing $a(\zeta) + b(\zeta)$ over $\mathrm{GF}[2]$, i.e., for all $i \in \{0, \ldots, n - 1\}$, the ith coefficient of $a(\zeta) + b(\zeta)$ is $a_i \oplus b_i$. Computing $a \cdot b$ over $\mathrm{GF}[2^n]$ consists of computing $a(\zeta) \cdot b(\zeta) \bmod r(\zeta)$, where $a(\zeta) \cdot b(\zeta)$ is polynomial multiplication over $\mathrm{GF}[2]$ that results in a polynomial of degree $2n-2$, and $r(\zeta)$ is a fixed irreducible polynomial of degree n. The zero element of $\mathrm{GF}[2^n]$ is the identically zero polynomial with coefficients $a_i = 0$ for all $i \in \{0, \ldots, n - 1\}$, and the identity element is the polynomial with coefficients $a_0 = 1$ and $a_i = 0$ for all $i \in \{1, \ldots, n - 1\}$.

——————∞——————

In the Modulo Prime Space, $\langle X_0, \ldots, X_{p-1} \rangle$ are pairwise independent and the size of the space is p^2. We describe a way to construct a pairwise independent probability space for $\{0, 1\}$-valued random variables that has size linear in the number of random variables. This space will be used in the solution to the Vertex Partition problem described below, and it also plays a crucial role in some of the remaining lectures.

Inner Product Space : Let ℓ be a positive integer. The sample space is $\{0, 1\}^\ell$ and the distribution on sample points is $A \in_{\mathcal{U}} \{0, 1\}^\ell$. For all $i \in \{0, 1\}^\ell \setminus \{0^\ell\}$, define random variable

$$X_i(A) = A \odot i = \left(\sum_{j=1}^{\ell} A_j \cdot i_j \right) \bmod 2.$$

Claim : $\langle X_1, \ldots, X_{2^\ell-1} \rangle$ are uniformly distributed and pairwise independent.

Exercise 14 : Prove the pairwise independence property for the Inner Product Space. ♠

Witness Sampling Problem

Let $L : \{0,1\}^n \to \{0,1\}$ be an **RP** language and let $f : \{0,1\}^n \times \{0,1\}^{\ell(n)} \to \{0,1\}$ be the **P**-time function ensemble associated with L. Let $x \in \{0,1\}^n$ and let $Y \in_{\mathcal{U}} \{0,1\}^{\ell(n)}$. The function f has the property that if $x \in L$ then $\Pr_Y[f(x,Y) = 1] \geq \epsilon$ for some fixed $\epsilon > 0$, and if $x \notin L$ then $\Pr_Y[f(x,Y) = 1] = 0$. We want to design a **P**-time function ensemble f' that has high probability of finding a witness when $x \in L$. More formally, $f'(x, \delta, Z)$ is a **P**-time function ensemble that on input $x \in \{0,1\}^n$ and $\delta > 0$ has the property that if $x \in L$ then

$$\Pr_Z[f(x, f'(x, \delta, Z)) = 1] \geq 1 - \delta.$$

We describe two methods to implement such an f':

Method 1 :

Let $m = \lceil \ln(1/\delta)/\epsilon \rceil$.

Independently choose $y_1, \ldots, y_m \in_{\mathcal{U}} \{0,1\}^{\ell(n)}$.

For all $i \in \{1, \ldots, m\}$ do:

  If $f(x, y_i) = 1$ then output witness y_i and stop.

Output $0^{\ell(n)}$.

For the analysis, assume that $x \in L$. Let $Y_1, \ldots, Y_m \in_{\mathcal{U}} \{0,1\}^{\ell(n)}$. Because for each $i \in \{1, \ldots, m\}$, $\Pr_{Y_i}[f(x, Y_i) = 1] \geq \epsilon$,

$$\Pr_{\langle Y_1, \ldots, Y_m \rangle}[f(x, Y_i) = 0 \text{ for all } i \in \{1, \ldots, m\}] \leq (1 - \epsilon)^m \leq e^{-\epsilon m} \leq \delta.$$

This uses $\ell(n) \lceil \ln(1/\delta)/\epsilon \rceil$ random bits and $\lceil \ln(1/\delta)/\epsilon \rceil$ tests.

Method 2 :

Let $m = \lceil 1/(\epsilon\delta) \rceil$.

Let $S = \{\langle a, b \rangle : a, b \in \mathrm{GF}[2^{\ell(n)}]\}$.

Choose $\langle a, b \rangle \in_{\mathcal{U}} S$.

For all $i \in \{1, \ldots, m\}$ do:

   Compute $Y_i(a, b)$ as described in the Linear Polynomial Space.

   If $f(x, Y_i(a, b)) = 1$ then output witness $Y_i(a, b)$ and stop.

Output $0^{\ell(n)}$.

For the analysis, assume that $x \in L$. The total number of witness tests this algorithm performs is m, and we can assume that $m < 2^{\ell(n)}$ because with $2^{\ell(n)}$ tests a witness can be found by exhaustive search of all strings in $\{0, 1\}^{\ell(n)}$. Let $\langle A, B \rangle \in_{\mathcal{U}} S$. By the properties of the Linear Polynomial Space, $Y_1(A, B), \ldots, Y_m(A, B)$ are uniformly distributed in $\{0, 1\}^{\ell(n)}$ and pairwise independent. Let

$$\alpha = \Pr_{A, B}[f(x, Y_1(A, B)) = 1] = \cdots = \Pr_{A, B}[f(x, Y_m(A, B)) = 1] \geq \epsilon.$$

Define random variable

$$Z(A, B) = 1/m \cdot \sum_{i=1}^{m} (f(x, Y_i(A, B)) - \alpha).$$

The only way that all m possible witnesses fail to be witnesses is if $Z(A, B) = -\alpha$, and this can only happen if $|Z(A, B)| \geq \alpha$. But,

$$\Pr_{A, B}[|Z(A, B)| \geq \alpha] \leq \mathrm{E}_{A, B}[Z(A, B)^2]/\alpha^2$$

$$= m\alpha(1 - \alpha)/(m^2 \alpha^2) \leq 1/(m\alpha) \leq \delta.$$

This follows by first applying Chebychev's Inequality (see Exercise 5 on page 12), then using the fact that, for $i \neq j$, $\mathrm{E}_{A, B}[(f(x, Y_i(A, B)) - \alpha)(f(x, Y_j(A, B)) - \alpha)] = 0$ (which uses the pairwise independence property) and that $\mathrm{E}_{A, B}[(f(x, Y_i(A, B)) - \alpha)^2] = \alpha(1 - \alpha)$, and finally using the fact that $\alpha \geq \epsilon$ and $m \geq 1/(\epsilon\delta)$. It follows that at least one of the $Y_1(a, b), \ldots, Y_m(a, b)$ will be a witness with probability at least $1 - \delta$.

This uses $2\ell(n)$ random bits and $\lceil 1/(\epsilon\delta) \rceil$ tests. Compared to the first method, fewer random bits are used at the expense of more tests.

The following exercise shows that there is a smooth transition between the two methods just described.

Exercise 15 : Describe a natural hybrid between the two methods that uses $2k\ell(n)$ random bits and

$$m = \max\left\{ \lceil 2/\epsilon \rceil \cdot \left\lceil (1/\delta)^{1/k} \right\rceil, \lceil k/\epsilon \rceil \right\}$$

witness tests.

Hint : Use Exercise 13 (page 57) to produce a $2k$-wise independent distribution on random variables using $2k\ell(n)$ random bits in total. Generalize Chebychev's inequality to show that when the random variables are $2k$-wise independent then

$$\Pr[|Z| \geq \alpha] \leq E[Z^{2k}]/\alpha^{2k}.$$

Show that

$$E[Z^{2k}] \leq (m\alpha + k)^k / m^{2k}.$$

♠

When $k = \lceil \log(1/\delta) \rceil$ then the hybrid method is better than Method 1, i.e., the number of random bits is smaller by a factor of ϵ whereas the number of witness tests is essentially the same, i.e.,

$$\lceil \log(1/\delta)/\epsilon \rceil .$$

Vertex Partition Problem

Input : An undirected graph $G = (V, E)$, $\#V = n$, $\#E = m$. For each edge $e \in E$ a weight $\text{wt}(e) \in \mathcal{Z}^+$ is specified.

Output : A vertex partition $\langle V_0, V_1 \rangle$ of V with the property that at least $1/2$ of the total edge weight crosses the partition. In other words, define $W = \sum_{e \in E} \text{wt}(e)$ and define the weight of $\langle V_0, V_1 \rangle$ as

$$W(V_0, V_1) = \sum_{\{e = (v_0, v_1): v_0 \in V_0, v_1 \in V_1\}} \text{wt}(e).$$

The output is a vertex partition $\langle V_0, V_1 \rangle$ such that $W(V_0, V_1) \geq 1/2 \cdot W$.

We use the Inner Product Space (page 58) to solve this problem in polynomial time. Although more efficient solutions exist, this solution illustrates in a simple setting the general idea of exploiting both the randomness and structural properties of the space. The analysis shows that some point in the space defines a good partition, and because the space is small and easy to construct, a good partition can be found by exhaustive search of the space.

Let $\ell = \lceil \log(n+1) \rceil$. For each vertex $i \in \{0,1\}^\ell \setminus \{0^\ell\}$ and for each sample point $a \in \{0,1\}^\ell$, define $X_i(a) = i \odot a$. Let

$$V_0(a) = \{i \in \{1, \ldots, n\} : X_i(a) = 0\}$$

and

$$V_1(a) = \{i \in \{1, \ldots, n\} : X_i(a) = 1\}.$$

Let $A \in_\mathcal{U} \{0,1\}^\ell$. By the pairwise independence property, for all pairs $i, j \in V$, $i \neq j$,

$$\Pr_A[X_i(A) \neq X_j(A)] = 1/2.$$

Thus, for each edge e, the probability that the two endpoints of e are on opposite sides of the partition is exactly $1/2$. From this it follows that

$$E_A[W(V_0(A), V_1(A))] = 1/2 \cdot W.$$

Since the average value of the weight crossing the partition with respect to A is $1/2 \cdot W$, there is at least one $a \in \{0,1\}^\ell$ where $W(V_0(a), V_1(a)) \geq 1/2 \cdot W$. The vertex partition that maximizes $W(V_0(a), V_1(a))$ over all $a \in \{0,1\}^\ell$ is a solution to the problem.

For each sample point it takes time $\mathcal{O}(m + n \cdot \lceil \log(n) \rceil)$ to compute the weight, and there are $\mathcal{O}(n)$ sample points to be checked. Thus, the total running time is $\mathcal{O}(n(m + n \cdot \lceil \log(n) \rceil))$. The resulting algorithm is deterministic even though the analysis is probabilistic.

Exercise 16 : Find an $\mathcal{O}(m+n)$ time algorithm for the Vertex Partition Problem.

Hint : Don't think about the method described above.   ♠

Exercise 17 : Find a parallel algorithm that uses $\mathcal{O}(m+n)$ processors and runs in time $\mathcal{O}(\log^2(m + n))$ for the Vertex Partition Problem.

Hint : Find a good sample point $a \in \{0,1\}^\ell$ by determining the bits $a_1, \ldots, a_\ell$ in sequence one bit at a time.   ♠

Exercise 18 : Let p be a positive integer and let $X_1, \ldots, X_n \in_\mathcal{U} \mathcal{Z}_p$ be a sequence of four-wise independent random variables. Define random variable

$$Y = \min\{|X_i - X_j| : 1 \leq i < j \leq n\}.$$

Prove that there is a constant $c > 0$ such that for any $\alpha \leq 1$

$$\Pr[Y \leq \alpha p/n^2] \geq c\alpha.$$

Hint : Let S be the set of $n(n-1)/2$ unordered pairs $\{(i,j) : 1 \le i < j \le n\}$. For fixed α, consider the sequence of $\{0,1\}$-valued random variables $\{Z_s : s \in S\}$, where if $s = (i,j)$ then $Z_s = 1$ if $|X_i - X_j| \le \alpha p/n^2$ and $Z_s = 0$ otherwise. Using the first two terms of the inclusion-exclusion formula, show that for any α, $\Pr[\exists s \in S : Z_s = 1] \ge \sum_{s \in S} \Pr[Z_s = 1] - \sum_{s,t \in S, s \ne t} \Pr[Z_s = 1 \wedge Z_t = 1]$. ♠

Lecture 6

Overview

We give the definition of the inner product bit for a function and define what it means for this bit to be hidden. We prove that the inner product bit is hidden for a one-way function. One immediate application is a simple construction of a pseudorandom generator from any one-way permutation.

The inner product bit is hidden for a one-way function

In this lecture, we introduce and prove the Hidden Bit Theorem. There are several technical parts in the reduction from any one-way function f to a pseudorandom generator g. Intuitively, the Hidden Bit Theorem is the part that transforms the one-wayness of f into a bit b such that: (1) b is completely determined by information that is available to any adversary; (2) nevertheless b looks random to any appropriately time-restricted adversary. Intuitively, it is from this bit b that the generator g eventually derives its pseudorandomness. The guarantee from the reduction is that any successful adversary for distinguishing the output of g from a truly random string can be converted into an adversary for predicting b, which in turn can be converted into an adversary for inverting f.

Definition (inner product bit is hidden): Let $f : \{0,1\}^n \to \{0,1\}^{\ell(n)}$ be a **P**-time function ensemble, where the input is private, and thus the security parameter is n. Let $z \in \{0,1\}^n$. Define the inner product bit of f with respect to z to be $x \odot z$. Let $X \in_{\mathcal{U}} \{0,1\}^n$ and let $Z \in_{\mathcal{U}} \{0,1\}^n$. Let $A : \{0,1\}^{\ell(n)} \times \{0,1\}^n \to \{0,1\}$ be an adversary. The success probability (prediction probability) of A for the inner product bit of f is

$$\delta(n) = |\Pr_{X,Z}[A(f(X), Z) = X \odot Z] - \Pr_{X,Z}[A(f(X), Z) \neq X \odot Z]|.$$

Then, the *inner product bit of f is a $\mathbf{S}(n)$-secure* if every adversary has time-success ratio at least $\mathbf{S}(n)$. ♣

The heart of the proof of Hidden Bit Theorem is the following technical theorem. We prove this theorem after first using it to prove the Hidden Bit Theorem.

Hidden Bit Technical Theorem : Let $B : \{0,1\}^n \rightarrow \{0,1\}$ be a function ensemble. Let $Z \in_{\mathcal{U}} \{0,1\}^n$ and for each $x \in \{0,1\}^n$ define

$$\delta_x^B = \Pr_Z[B(Z) = x \odot Z] - \Pr_Z[B(Z) \neq x \odot Z].$$

There is an oracle adversary S such that for any B, S^B on input $\delta > 0$ produces a list $\mathcal{L} \subseteq \{0,1\}^n$ with the following property: For all $x \in \{0,1\}^n$, if $\delta_x^B \geq \delta$ then $x \in \mathcal{L}$ with probability at least $1/2$, where this probability is with respect to the random bits used by oracle adversary S^B. The running time of S^B is $\mathcal{O}(n^3 T/\delta^4)$, where T is the running time of B.

Hidden Bit Theorem : If f is a one-way function then the inner product bit of f is hidden. The reduction is poly-preserving.

PROOF: Suppose there is an adversary A for the inner product bit of f with success probability $\delta(n)$ and run time $T(n)$. We describe an oracle adversary S such that S^A is an adversary for f as a one-way function.

Let $Z \in_{\mathcal{U}} \{0,1\}^n$ and for $x \in \{0,1\}^n$ define

$$\delta_x^A = \Pr_Z[A(f(x), Z) = x \odot Z] - \Pr_Z[A(f(x), Z) \neq (x \odot Z)].$$

Let $X \in_{\mathcal{U}} \{0,1\}^n$. Because, for any $x \in \{0,1\}^n$, $|\delta_x^A| \leq 1$ and because $\mathrm{E}_X[\delta_X^A] = \delta(n)$, it follows that $\Pr_X[\delta_X^A \geq \delta(n)/2] \geq \delta(n)/2$. The oracle adversary S we describe below has the property that if $\delta_x^A \geq \delta(n)/2$ then S^A on input $f(x)$ succeeds in producing an x' such that $f(x') = f(x)$ with probability at least $1/2$. From this it follows that the inverting probability of S^A for f is at least $\delta(n)/4$.

Suppose the input to S^A is $f(x)$, where $\delta_x^A \geq \delta(n)/2$. Let S' be the oracle adversary described in the Hidden Bit Technical Theorem and let $B(z) = A(f(x), z)$. The first step of S^A is to run S'^B with input $\delta = \delta(n)/2$. When S' makes an oracle query to B with input z, S runs A on input $\langle f(x), z \rangle$ and returns the answer $B(z) = A(f(x), z)$ to S'. Because $\delta_x^A \geq \delta(n)/2$, by the Hidden Bit Technical Theorem, x is in the list $\mathcal{L}$ produced by S'^B with probability at least $1/2$. The final step of S^A is to do the following for all $x' \in \mathcal{L}$: Compute $f(x')$ and if $f(x') = f(x)$ then output x'.

The success probability of S^A for inverting $f(X)$ is at least $\delta(n)/4$. From the Hidden Bit Technical Theorem, it is not hard to see that the running time of S^A is dominated by the running time of S' making queries to A to produce the list $\mathcal{L}$, which is $\mathcal{O}(n^3 T(n)/\delta(n)^4)$, where $T(n)$ is the running time of A. Thus, the time-success ratio of S^A is $\mathcal{O}(n^3 T(n)/\delta(n)^5)$. ∎

The following exercise shows that an immediate application of the Hidden Bit Theorem is the construction of a pseudorandom generator from a one-way permutation.

Exercise 19 :   From the Hidden Bit Theorem, show that if $f(x)$ is a one-way permutation then $g(x, z) = \langle f(x), x \odot z \rangle$ is a pseudorandom generator. The reduction should be poly-preserving.   ♠

The following exercise shows that the inner product bit is special, i.e., it is certainly not the case that any bit of the input to f is hidden if f is a one-way function.

Exercise 20 :   Let $X \in \{0, 1\}^n$. Describe a one-way permutation $f : \{0, 1\}^n \rightarrow \{0, 1\}^{\ell(n)}$ where X_1 is not hidden given $f(X)$. Let $f : \{0, 1\}^n \rightarrow \{0, 1\}^{\ell(n)}$ be a **P**-time function ensemble and let $I \in_{\mathcal{U}} \{1, \ldots, n\}$. Show that if X_I can be predicted with probability greater than $1 - 1/(2n)$ given $f(X)$ then f is not a one-way function.   ♠

The converse of the Hidden Bit Theorem is not true, i.e., there is a function f where the inner product bit is hidden but f is not a one-way function. This is the point of the following exercise.

Exercise 21 :   Describe a **P**-time function ensemble $f : \{0, 1\}^n \rightarrow \{0, 1\}^{\ell(n)}$ which is certainly not a one-way function but for which the inner product bit is provably 2^n-secure.   ♠

Generalized Inner Product Space

The proof of the Hidden Bit Technical Theorem uses the paradigm discussed in Lecture 5. For this, we use the following generalization of the Inner Product Space (page 58).

Generalized Inner Product Space :   Let $\ell = \lceil \log(m + 1) \rceil$. The sample space is $\{0, 1\}^{n \times \ell}$ and the distribution on sample points is $V \in_{\mathcal{U}} \{0, 1\}^{n \times \ell}$. For all $j \in \{0, 1\}^\ell$, define random variable

$$T_j(V) = V \odot j.$$

It can be verified that $\langle T_1(V), \ldots, T_m(V) \rangle$ are uniformly distributed on $\{0, 1\}^n$ and pairwise independent.

Proof of the Hidden Bit Technical Theorem

For the proof of the Hidden Bit Technical Theorem, we find it convenient to consider bits as being $\{1, -1\}$-valued instead of $\{0, 1\}$-valued. This notation is described on page 4.

PROOF of the Hidden Bit Technical Theorem: Fix $x \in \{0, 1\}^n$ such that $\delta_x^B \geq \delta$. Let $Z \in_{\mathcal{U}} \{0, 1\}^n$. In the $\{1, -1\}$ notation, we can write

$$\delta_x^B = \mathrm{E}_Z \left[\overline{B(Z)} \cdot \overline{x \odot Z} \right].$$

For all $i = 1, \ldots, n$, let $e_i \in \{0, 1\}^n$ be the bit string $\langle 0^{i-1}, 1, 0^{n-i} \rangle$ and let

$$\mu_i = \delta_x^B \cdot \overline{x}_i.$$

Since, for any $z \in \{0, 1\}^n$,

$$\overline{x \odot (e_i \oplus z)} = \overline{x \odot z} \cdot \overline{x}_i,$$

it follows that

$$\overline{B(z)} \cdot \overline{x \odot (e_i \oplus z)} = \overline{B(z)} \cdot \overline{x \odot z} \cdot \overline{x}_i,$$

and thus

$$\mathrm{E}_Z \left[\overline{B(Z)} \cdot \overline{x \odot (e_i \oplus Z)} \right] = \mathrm{E}_Z \left[\overline{B(Z)} \cdot \overline{x \odot Z} \right] \cdot \overline{x}_i = \delta_x^B \cdot \overline{x}_i = \mu_i.$$

Setting $Z' = e_i \oplus Z$ it is easy to see that $Z' \in_{\mathcal{U}} \{0, 1\}^n$ and $Z = e_i \oplus Z'$. Thus,

$$\mathrm{E}_{Z'} \left[\overline{B(e_i \oplus Z')} \cdot \overline{x \odot Z'} \right] = \mu_i.$$

The idea is to compute, simultaneously for all $i \in \{1, \ldots, n\}$, a good approximation Y_i of μ_i. We say that Y_i is a good approximation if $|Y_i - \mu_i| < \delta$. Define

$$\mathrm{bit}(Y_i) = \begin{cases} 0 & \text{if } Y_i \geq 0 \\ 1 & \text{if } Y_i < 0 \end{cases}$$

Because $|\mu_i| \geq \delta$, if Y_i is a good approximation then $\mathrm{bit}(Y_i) = x_i$. Let $m = \lceil 2n/\delta^2 \rceil$ and let $T_1, \ldots, T_m \in_{\mathcal{U}} \{0, 1\}^n$ be pairwise independent random variables. Let

$$Y_i = 1/m \cdot \sum_{j=1}^{m} \overline{B(e_i \oplus T_j)} \cdot \overline{x \odot T_j}.$$

Then, using the pairwise independence of the random variables and the fact that, for all j,

$$\mathrm{E} \left[\left(\overline{B(e_i \oplus T_j)} \cdot \overline{x \odot T_j} - \mu_i \right)^2 \right] \leq 1,$$

(this fact takes a bit of justification) it follows that

$$E[(Y_i - \mu_i)^2] \leq 1/m.$$

From Chebychev's inequality it then follows that

$$\Pr\left[|Y_i - \mu_i| \geq \delta\right] \leq E[(Y_i - \mu_i)^2]/\delta^2 \leq 1/(m\delta^2) \leq 1/(2n).$$

From this it follows that

$$\Pr\left[\exists i \in \{1, \ldots, n\} : |Y_i - \mu_i| \geq \delta\right] \leq 1/2,$$

and so

$$\Pr\left[\text{ for all } i \in \{1, \ldots, n\} : |Y_i - \mu_i| < \delta\right] \geq 1/2. \tag{3}$$

The only remaining difficulty is how to compute Y_i given $T_1, \ldots, T_m$. Everything is relatively easy to compute, except for the values of $x \odot T_j$ for all $j \in \{1, \ldots, m\}$. If $T_1, \ldots, T_m$ are chosen in the obvious way, i.e., each is chosen independently of all the others, then we need to be able to compute $x \odot T_j$ correctly for all $j \in \{1, \ldots, m\}$ and there is probably no feasible way to do this. (Recall that we don't know the value of x.) Instead, the approach is to take advantage of the fact that the analysis only requires $T_1, \ldots, T_m$ to be pairwise independent.

Let $\ell = \lceil \log(m+1) \rceil$ and let $v \in \{0,1\}^{n \times \ell}$. Let $T_1(v), \ldots, T_m(v)$ be as described in the Generalized Inner Product Space (page 66), i.e., for all $v \in \{0,1\}^{n \times \ell}$ and for all $j \in \{0,1\}^{\ell} - 0^{\ell}$, $T_j(v) = v \odot j$. As we describe, this particular construction allows feasible enumeration of all possible values of $x \odot T_j(v)$ for all $j \in \{1, \ldots, m\}$ without knowing x. Because of the properties stated above,

$$x \odot T_j(v) = x \odot (v \odot j) = (x \odot v) \odot j.$$

Thus, it is easy to compute, for all $j \in \{1, \ldots, m\}$, the value of $x \odot T_j(v)$ given $x \odot v$. From this we can compute, for all $i \in \{1, \ldots, n\}$,

$$Y_i(v) = 1/m \cdot \sum_{j=1}^{m} \overline{B(e_i \oplus T_j(v))} \cdot \overline{(x \odot v) \odot j}.$$

There are only $2^{\ell} = \mathcal{O}(m)$ possible settings for $x \odot v$, and we try them all. For any x and v there is some $\beta \in \{0,1\}^{\ell}$ such that $\beta = x \odot v$. Let

$$Y_i(\beta, v) = 1/m \cdot \sum_{j=1}^{m} \overline{B(e_i \oplus T_j(v))} \cdot \overline{\beta \odot j},$$

i.e., $Y_i(\beta, v)$ is the value obtained when β is substituted for $x \odot v$ in the computation of $Y_i(v)$. Consider choosing $v \in_{\mathcal{U}} \{0,1\}^{n \times \ell}$. Since from equation (3) above, the probability that $Y_i(x \odot v, v)$ is a good approximation for all $i \in \{1, \ldots, n\}$ is at least one-half, it follows that with probability at least one-half there is at least one $\beta \in \{0,1\}^\ell$ such that $Y_i(\beta, v)$ is simultaneously a good approximation for all $i \in \{1, \ldots, n\}$. For this value of β and for such a v, $\langle \text{bit}(Y_1(\beta, v)), \ldots, \text{bit}(Y_n(\beta, v)) \rangle$ is equal to x.

Adversary S^B on input $\delta > 0$:

$m \leftarrow \lceil 2n/\delta^2 \rceil$.

$\ell \leftarrow \lceil \log(m+1) \rceil$.

$\mathcal{L} \leftarrow \emptyset$.

Choose $v \in_{\mathcal{U}} \{0,1\}^{n \times \ell}$.

For all $\beta \in \{0,1\}^\ell$ do:

    For all $j = 1, \ldots, m$ do:

    Compute $T_j(v) = v \odot j$.

    For all $i = 1, \ldots, n$ do:

    Compute $Y_i(\beta, v) = 1/m \cdot \sum_{j=1}^m \overline{B(e_i \oplus T_j(v))} \cdot \overline{\beta \odot j}$.

    $\mathcal{L} \leftarrow \mathcal{L} \cup \{\langle \text{bit}(Y_1(\beta, v)), \ldots, \text{bit}(Y_n(\beta, v)) \rangle\}$.

From the above analysis, it follows that $x \in \mathcal{L}$ with probability at least $1/2$, where this probability is over the random choice of v.

As long as the running time T for computing B is large compared to n (which it is in our use of the Hidden Bit Technical Theorem to prove the Hidden Bit Theorem), the running time of S^B is $\mathcal{O}\left(n^3 T/\delta^4\right)$. ∎

Lecture 7

Overview

We describe statistical measures of distance between probability distributions and define what it means for two distributions to be computationally indistinguishable. We prove that many inner product bit are hidden for a one-way function.

Measures of distance between probability distributions

For the following definitions and exercises, let $\mathcal{D}_n : \{0,1\}^n$ and $\mathcal{E}_n : \{0,1\}^n$ be distributions, and let $X \in_{\mathcal{D}_n} \{0,1\}^n$ and $Y \in_{\mathcal{E}_n} \{0,1\}^n$.

Definition (statistically distinguishable): The *statistical distance* between $\mathcal{D}_n$ and $\mathcal{E}_n$ is

$$\text{dist}(\mathcal{D}_n, \mathcal{E}_n) = 1/2 \cdot \sum_{z \in \{0,1\}^n} |\Pr_X[X = z] - \Pr_Y[Y = z]|.$$

Equivalently,

$$\text{dist}(\mathcal{D}_n, \mathcal{E}_n) = \max\left\{\Pr_X[X \in S] - \Pr_Y[Y \in S] : S \subseteq \{0,1\}^n\right\}.$$

We use $\text{dist}(X, Y)$ and $\text{dist}(\mathcal{D}_n, \mathcal{E}_n)$ interchangeably. We say $\mathcal{D}_n$ and $\mathcal{E}_n$ are at most $\epsilon(n)$-statistically distinguishable if $\text{dist}(\mathcal{D}_n, \mathcal{E}_n) \leq \epsilon(n)$. ♣

Definition (statistical test): A *statistical test* t for $\mathcal{D}_n$ and $\mathcal{E}_n$ is a function $t : \{0,1\}^n \rightarrow \{0,1\}$. The success probability (distinguishing probability) of t for $\mathcal{D}_n$ and $\mathcal{E}_n$ is

$$\delta(n) = |\Pr_X[t(X) = 1] - \Pr_Y[t(Y) = 1]|.$$

♣

Exercise 22 : Let $f : \{0,1\}^n \rightarrow \{0,1\}^{\ell(n)}$ be a function ensemble. Show that $\text{dist}(f(X), f(Y)) \leq \text{dist}(X, Y)$. ♠

Exercise 22 implies that the distinguishing probability of any statistical test for $\mathcal{D}_n$ and $\mathcal{E}_n$ is at most $\text{dist}(\mathcal{D}_n, \mathcal{E}_n)$. The following exercise shows there is a statistical test that achieves this maximum.

Exercise 23 : Describe a statistical test t such that $\delta(n) = \text{dist}(\mathcal{D}_n, \mathcal{E}_n)$. ♠

Exercise 24 : Prove that for any triple of distributions $\mathcal{D}_n^1 : \{0,1\}^n$, $\mathcal{D}_n^2 : \{0,1\}^n$, and $\mathcal{D}_n^3 : \{0,1\}^n$,

$$\text{dist}(\mathcal{D}_n^1, \mathcal{D}_n^3) \leq \text{dist}(\mathcal{D}_n^1, \mathcal{D}_n^2) + \text{dist}(\mathcal{D}_n^2, \mathcal{D}_n^3).$$

♠

Exercise 25 : Let $f : \{0,1\}^n \rightarrow \{0,1\}^{\ell(n)}$ be a function ensemble that can be computed in time $n^{\mathcal{O}(1)}$ on average, i.e., for $X \in_{\mathcal{U}} \{0,1\}^n$, $E_X[T(X)] = n^{\mathcal{O}(1)}$. where $T(x)$ is the time to compute f on input x. Show that for any $m(n) = n^{\mathcal{O}(1)}$ there is a $p(n) = n^{\mathcal{O}(1)}$ and a **P**-time function ensemble $f' : \{0,1\}^{p(n)} \rightarrow \{0,1\}^{\ell(n}$ such that $\text{dist}(f(X), f'(Z)) \leq 1/m(n)$, where $Z \in_{\mathcal{U}} \{0,1\}^{p(n)}$. ♠

Computationally limited tests

Exercises 22 and 23 together show for any pair of distributions there is a statistical test that achieves the maximum distinguishing probability possible. However, this test could have a large run time. A crucial idea in the development that follows is to limit the amount of computation time allowed for a test.

Definition (computationally indistinguishable): Let

$$\mathcal{D}_n : \{0,1\}^n \rightarrow \{0,1\}^{\ell(n)}$$

and

$$\mathcal{E}_n : \{0,1\}^n \rightarrow \{0,1\}^{\ell(n)}$$

be probability ensembles with common security parameter n. Let $X \in_{\mathcal{D}_n} \{0,1\}^{\ell(n)}$ and $Y \in_{\mathcal{E}_n} \{0,1\}^{\ell(n)}$. Let $A : \{0,1\}^{\ell(n)} \rightarrow \{0,1\}$ be an adversary. The success probability (distinguishing probability) of A for $\mathcal{D}_n$ and $\mathcal{E}_n$ is

$$\delta(n) = |\Pr_X[A(X) = 1] - \Pr_Y[A(Y) = 1]|.$$

We say $\mathcal{D}_n$ and $\mathcal{E}_n$ are $\mathbf{S}(n)$-*secure computationally indistinguishable* if every adversary has time-success ratio at least $\mathbf{S}(n)$. ♣

Exercise 26 : Let $\mathcal{D}_n : \{0,1\}^n \rightarrow \{0,1\}^{\ell(n)}$ and $\mathcal{E}_n : \{0,1\}^n \rightarrow \{0,1\}^{\ell(n)}$ be probability ensembles with common security parameter n. Prove that if $\mathcal{D}_n$ and $\mathcal{E}_n$ are at most $\epsilon(n)$-statistically distinguishable then $\mathcal{D}_n$ and $\mathcal{E}_n$ are $(1/\epsilon(n))$-secure computationally indistinguishable. ♠

Exercise 27 : Let $\mathcal{D}_n^1 : \{0,1\}^n \rightarrow \{0,1\}^{\ell(n)}$, $\mathcal{D}_n^2 : \{0,1\}^n \rightarrow \{0,1\}^{\ell(n)}$, and $\mathcal{D}_n^3 : \{0,1\}^n \rightarrow \{0,1\}^{\ell(n)}$ be probability ensembles with common security parameter n. Prove that if $\mathcal{D}_n^1$ and $\mathcal{D}_n^2$ are $\mathbf{S}_{12}(n)$)-secure computationally indistinguishable and $\mathcal{D}_n^2$ and $\mathcal{D}_n^3$ are $\mathbf{S}_{23}(n)$)-secure computationally indistinguishable then $\mathcal{D}_n^1$ and $\mathcal{D}_n^3$ are $\mathbf{S}_{13}(n)$)-secure computationally indistinguishable, where

$$\mathbf{S}_{13}(n) = \Omega(\min\{\mathbf{S}_{12}(n), \mathbf{S}_{23}(n)\}/n^{\mathcal{O}(1)}).$$

♠

We are often interested in probability ensembles that are **P**-samplable. (See page 6 for the definition.)

Exercise 28 :   Let $\mathcal{D}_n : \{0,1\}^n \rightarrow \{0,1\}^{\ell(n)}$ and $\mathcal{E}_n : \{0,1\}^n \rightarrow \{0,1\}^{\ell(n)}$ be **P**-samplable probability ensembles with common security parameter n. Let $k(n)$ be a polynomial parameter. Define **P**-samplable probability ensemble $\mathcal{D}_n' : \{0,1\}^{nk(n)} \rightarrow \{0,1\}^{n\ell(n)}$ and $\mathcal{E}_n' : \{0,1\}^{nk(n)} \rightarrow \{0,1\}^{n\ell(n)}$ with common security parameter $nk(n)$ as

$$\mathcal{D}_n' = \underbrace{\mathcal{D}_n \times \cdots \times \mathcal{D}_n}_{k(n)}$$

and

$$\mathcal{E}_n' = \underbrace{\mathcal{E}_n \times \cdots \times \mathcal{E}_n}_{k(n)}.$$

Describe an oracle adversary S such that if A is an adversary for $\mathcal{D}_n$ and $\mathcal{E}_n$ with time-success ratio $\mathbf{R}'(nk(n))$ then S^A is an adversary for $\mathcal{D}_n'$ and $\mathcal{E}_n'$ with time-success ratio $\mathbf{R}(n)$, where $\mathbf{R}(n) = n^{\mathcal{O}(1)} \cdot \mathcal{O}(\mathbf{R}'(nk(n)))$. ♠

Exercise 28 shows that if $\mathcal{D}_n$ and $\mathcal{E}_n$ are computationally indistinguishable then so are $\mathcal{D}_n'$ and $\mathcal{E}_n'$. It turns out to be crucial that both $\mathcal{D}_n$ and $\mathcal{E}_n$ are **P**-samplable for this reduction to be uniform. This exercise is a simple but crucial ingredient in the reduction from a one-way function to a pseudorandom generator. However, using many independent copies of a distribution in the reduction is the primary reason it is only weak-preserving. Another example of this kind of phenomena is the first reduction in Lecture 3.

Exercise 29 :   Let $\mathcal{D}_n : \{0,1\}^n \rightarrow \{0,1\}^{\ell(n)}$ and $\mathcal{E}_n : \{0,1\}^n \rightarrow \{0,1\}^{\ell(n)}$ be **P**-samplable probability ensembles with common security parameter n. Let $f : \{0,1\}^{\ell(n)} \rightarrow \{0,1\}^{p(n)}$ be a **P**-time function ensemble. Let $X \in_{\mathcal{D}_n} \{0,1\}^{\ell(n)}$ and $Y \in_{\mathcal{E}_n} \{0,1\}^{\ell(n)}$. Let $f(X)$ and $f(Y)$ be **P**-samplable probability ensembles with common security parameter n. Describe an oracle adversary S such that if A is an adversary for $f(X)$ and $f(Y)$ with time-success ratio $\mathbf{R}'(n)$ then S^A is an adversary for $\mathcal{D}_n$ and $\mathcal{E}_n$ with time-success ratio $\mathbf{R}(n)$, where $\mathbf{R}(n) = n^{\mathcal{O}(1)} \cdot \mathcal{O}(\mathbf{R}'(n))$. ♠

The following is a corollary of the Hidden Bit Theorem (page 65). It is essentially a restatement of the Hidden Bit Theorem in terms of two distributions being computationally indistinguishable. This shows how the one-wayness of a function f is converted into pseudorandomness via the inner product bit.

Construction of a hidden bit from a one-way function : Let $f : \{0,1\}^n \rightarrow \{0,1\}^{\ell(n)}$ be a one-way function. Let $X \in_{\mathcal{U}} \{0,1\}^n$, $Z \in \{0,1\}^n$ and $B \in_{\mathcal{U}} \{0,1\}$. Let $\mathcal{D}_n = \langle f(X), X \odot Z, Z \rangle$ and $\mathcal{E}_n = \langle f(X), B, Z \rangle$ be **P**-samplable probability ensembles with common security parameter n.

Hidden Bit Corollary : If f is a one-way function then $\mathcal{D}_n$ and $\mathcal{E}_n$ are computationally indistinguishable. The reduction is poly-preserving.

PROOF: Suppose there is an adversary $A : \{0,1\}^{\ell(n)+n+1} \rightarrow \{0,1\}$ that has success probability $\delta(n)$ for distinguishing $\mathcal{D}_n$ and $\mathcal{E}_n$. Without loss of generality, suppose A is more likely to output 1 when the input is fixed according to $X \odot Z$ then according to B, and thus the success probability of A is

$$\delta(n) = \Pr_{X,Z}[A(f(X), X \odot Z, Z) = 1] - \Pr_{X,Z,B}[A(f(X), B, Z) = 1].$$

We show there is an oracle adversary S such that

$$\Pr_{X,Z}[S^A(f(X), Z) = X \odot Z] - \Pr_{X,Z}[S^A(f(X), Z) \neq X \odot Z] = \delta(n),$$

where the running time of S^A is essentially the same as the running time of A. The proof then follows from the Hidden Bit Theorem (page 65). The input to S^A is $\langle y, z \rangle$, where $x \in_{\mathcal{U}} \{0,1\}^n$ and $y = f(x)$, and $z \in_{\mathcal{U}} \{0,1\}^n$.

Adversary S^A on input $\langle y, z \rangle$: .

Choose $u, v \in_{\mathcal{U}} \{0,1\}$.

If $A(y, u, z) = 1$ then output u else output v.

We show that S^A has the above claimed distinguishing probability. Let

$$
\begin{aligned}
C_0 &= \{\langle x, z \rangle : A(f(x), 0, z) = A(f(x), 1, z) = 0\}, \\
C_1 &= \{\langle x, z \rangle : A(f(x), 0, z) = A(f(x), 1, z) = 1\}, \\
C_{\neq,0} &= \{\langle x, z \rangle : A(f(x), 0, z) \neq A(f(x), 1, z) \wedge A(f(x), x \odot z, z) = 0\}, \\
C_{\neq,1} &= \{\langle x, z \rangle : A(f(x), 0, z) \neq A(f(x), 1, z) \wedge A(f(x), x \odot z, z) = 1\},
\end{aligned}
$$

and let

$$
\begin{aligned}
\delta_0 &= \Pr_{X,Z}[\langle X, Z \rangle \in C_0], \\
\delta_1 &= \Pr_{X,Z}[\langle X, Z \rangle \in C_1], \\
\delta_{\neq,0} &= \Pr_{X,Z}[\langle X, Z \rangle \in C_{\neq,0}], \\
\delta_{\neq,1} &= \Pr_{X,Z}[\langle X, Z \rangle \in C_{\neq,1}].
\end{aligned}
$$

It is easy to verify that

$$
\delta(n) = (\delta_1 + \delta_{\neq,1}) - (\delta_1 + 1/2 \cdot (\delta_{\neq,0} + \delta_{\neq,1})) = 1/2 \cdot (\delta_{\neq,1} - \delta_{\neq,0}).
$$

Consider the behavior of $S^A(f(x), z)$ with respect to fixed values for x and z and random values for u and v. If $\langle x, z \rangle \in C_0$ then S^A always outputs v, and if $\langle x, z \rangle \in C_1$ then S^A always outputs u. In either case, the probability the output is equal to $x \odot z$ minus the probability the output is not equal to $x \odot z$ is 0. If $\langle x, z \rangle \in C_{\neq,1}$ then, with probability $1/2$, $u = x \odot z$ and u is output, and, with probability $1/2$, $u \neq x \odot z$ and v is output. Overall the output is equal to $x \odot z$ with probability $3/4$, and thus the probability the output is equal to $x \odot z$ minus the probability the output is not equal to $x \odot z$ is $1/2$. If $\langle x, z \rangle \in C_{\neq,0}$ then, using similar reasoning, the probability the output is equal to $x \odot z$ minus the probability the output is not equal to $x \odot z$ is $-1/2$. Thus,

$$
\Pr_{X,Z}[S^A(f(X), Z) = X \odot Z] - \Pr_{X,Z}[S^A(f(X), Z) \neq X \odot Z]
$$

is equal to $1/2 \cdot (\delta_{\neq,1} - \delta_{\neq,0}) = \delta(n)$.   ■

---

one-way permutation → a stretching pseudorandom generator

Combining the Hidden Bit Corollary (page 73) and the Stretching Theorem (page 52) immediately yields a construction of a pseudorandom generator that stretches by an arbitrary polynomial amount based on any one-way permutation.

Construction of pseudorandom generator from a one-way permutation :   Let $f : \{0,1\}^n \rightarrow \{0,1\}^n$ be a one-way permutation. Let $\ell(n) > n$ and $\ell(n) = n^{\mathcal{O}(1)}$. Define **P**-time function ensemble $g : \{0,1\}^n \times \{0,1\}^n \rightarrow \{0,1\}^{\ell(n)}$ as

$$
g(x, z) = \langle x \odot z, f(x) \odot z, f^{(2)}(x) \odot z, \ldots, f^{(\ell(n))}(x) \odot z, z \rangle,
$$

where $f^{(i)}$ is the function f composed with itself i times. The first input x is private and the second input z is public.

Theorem 7.1 : If f is a one-way permutation then g is a pseudorandom generator. The reduction from f to g is poly-preserving.

The following exercise is similar to Exercise 19 (page 66).

Exercise 30 : Prove Theorem 7.1.   ♠

Many Hidden Bits

We generalize the Hidden Bit Corollary (page 73) to the Many Hidden Bits Theorem below.

Construction of many hidden bits from a one-way function : Let $f : \{0,1\}^n \to \{0,1\}^{\ell(n)}$ be a P-time function ensemble. Let $X \in_{\mathcal{U}} \{0,1\}^n$. Let $r(n)$ be a positive integer valued function, let $Z \in_{\mathcal{U}} \{0,1\}^{n \times r(n)}$ and let $B \in_{\mathcal{U}} \{0,1\}^{r(n)}$. Let $\mathcal{D}_n = \langle f(X), X \odot Z, Z \rangle$ and $\mathcal{E}_n = \langle f(X), B, Z \rangle$.

———————∞———————

The proof of the following theorem from the Many Hidden Bits Technical Theorem is analogous to the proof of the Hidden Bit Theorem (page 65) from the Hidden Bit Technical Theorem (page 65), and is omitted.

Many Hidden Bits Theorem : If f is a one-way function then $\mathcal{D}_n$ and $\mathcal{E}_n$ are computationally indistinguishable. The reduction is poly-preserving when $r(n)$ is set to the logarithm of the security of f. ■

Many Hidden Bits Technical Theorem : Let n and r be positive integers, let $Z \in_{\mathcal{U}} \{0,1\}^{n \times r}$, and $B \in_{\mathcal{U}} \{0,1\}^r$. Let $A : \{0,1\}^r \times \{0,1\}^n \to \{0,1\}$ be an adversary. For all $x \in \{0,1\}^n$ define

$$\delta_x = \Pr_Z[A(x \odot Z, Z) = 1] - \Pr_{B,Z}[A(B, Z) = 1].$$

There is an oracle adversary S that on input δ produces a list $L \subseteq \{0,1\}^n$ with the following properties when making oracle queries to A. For all $x \in \{0,1\}^n$, if $\delta_x \geq \delta$ then the probability that $x \in L$ is at least $1/2$, where this probability is with respect to the random bits used by oracle adversary S^A. The running time of S^A is

$$(2^r T/\delta)^{\mathcal{O}(1)},$$

where T is the running time of A.

PROOF: The description of oracle adversary S depends on the following two intermediary oracle adversaries. These oracle adversaries convert the advantage of A for guessing the inner product of x with r random column vectors, each of length n, into an advantage for guessing the inner product of x with a single column vector of length n. We use the Hidden Bit Technical Theorem (page 65) to complete the proof.

We first describe an oracle adversary M' that uses A to gain an advantage in predicting $x \odot z$ given $z \in \{0,1\}^{n \times r}$.

Adversary M'^A on input z : .

Choose $u, v \in_{\mathcal{U}} \{0,1\}^r$.

If $A(u, z) = 1$ then output u else output v.

The run time of M'^A is the run time of A plus $n^{\mathcal{O}(1)}$.

Lemma 1 : Let $Z \in_{\mathcal{U}} \{0,1\}^{n \times r}$. For all $x \in \{0,1\}^n$,

$$\Pr_Z[M'^A(Z) = x \odot Z] = (1 + \delta_x)/2^r.$$

PROOF: Fix $x \in \{0,1\}^n$, let $U \in_{\mathcal{U}} \{0,1\}^r$ and $V \in_{\mathcal{U}} \{0,1\}^r$. For all $z \in \{0,1\}^{n \times r}$ let

$$\epsilon(z) = \Pr_U[A(U, z) = 1].$$

Then,

$$\delta_x = \mathrm{E}_Z[A(x \odot Z, Z)] - \mathrm{E}_Z[\epsilon(Z)].$$

We see how well the output of M'^A predicts $x \odot z$ for a fixed value of z with respect to U and V.

- z is such that $A(x \odot z, z) = 1$.

  - With probability 2^{-r}: $U = x \odot z$ and the output is correct.
  - With probability $(1 - \epsilon(z))$: $A(U, z) = 0$ and the output V is correct with probability 2^{-r}

- z is such that $A(x \odot z, z) = 0$.

  - With probability $(1 - \epsilon(z))$: $A(U, z) = 0$ and the output V is correct with probability 2^{-r}.

Putting this together, the probability the output is correct is

$$2^{-r} \cdot (\mathrm{E}_Z[A(x \odot Z, Z)] + \mathrm{E}_Z[1 - \epsilon(Z)]) = (1 + \delta_x)/2^r.$$

This completes the proof of Lemma 1. ∎

We now describe an oracle adversary M that makes one query to M'^A. We can view M^A as an oracle adversary that makes exactly one query to A (indirectly, via M'^A). The input to M^A is $y \in_\mathcal{U} \{0,1\}^n$.

Adversary M^A on input y : .

Choose $i \in_\mathcal{U} \{0,1\}^r \setminus \{0^r\}$

Let $\ell = \min\{j : i_j = 1\}$.

Choose $z \in_\mathcal{U} \{0,1\}^{n \times r}$.

Let $z' \in \{0,1\}^{n \times r}$ be z with the ℓth column, z_ℓ, replaced by $z_\ell \oplus (z \odot i) \oplus y$.

Output $M'^A(z') \odot i$.

The run time of M^A is the run time for the query to A plus $n^{\mathcal{O}(1)}$. Let $Y \in_\mathcal{U} \{0,1\}^n$, $I \in_\mathcal{U} \{0,1\}^r \setminus \{0^r\}$ and $Z \in_\mathcal{U} \{0,1\}^{n \times r}$. Define

$$\delta'_x = \mathrm{E}_{Y,I,Z}\left[\overline{M^A(Y)} \cdot \overline{(x \odot Y)}\right],$$

i.e., δ'_x is the correlation between the $x \odot Y$ and $M^A(Y)$.

Lemma 2 : For all $x \in \{0,1\}^n$, $\delta'_x = \delta_x/(2^r - 1)$.

PROOF: Because $Y \in_\mathcal{U} \{0,1\}^n$, Z' is uniformly distributed and Z' and I are independent. Furthermore,

$$Z' \odot I = (Z \odot I) \oplus (Z \odot I) \oplus Y = Y.$$

Lemma 1 shows the following with respect to Z'.

- With probability $(1 + \delta_x)/2^r$: $M'^A(Z') = x \odot Z'$. The output is $x \odot Z' \odot I = x \odot Y$, and thus the correlation is 1.

- With probability $1 - (1 + \delta_x)/2^r$: $M'^A(Z') \neq x \odot Z'$. Fix z' so that $M'^A(z') \neq x \odot z'$ and let $b_0 = M'^A(z')$ and $b_1 = x \odot z'$. Let $J \in_\mathcal{U} \{0,1\}^r$. From the pairwise independence property of the Inner Product Space (page 58), it follows that

$$\mathrm{E}_J\left[\overline{(b_0 \odot J)} \cdot \overline{(b_1 \odot J)}\right] = 0.$$

Note that if $j = 0^r$ then $\overline{(b_0 \odot j)} \cdot \overline{(b_1 \odot j)} = 1$. From this it follows that the correlation is

$$\mathbf{E}_I\left[\overline{(b_0 \odot I)} \cdot \overline{(b_1 \odot I)}\right] = \frac{-1}{2^r - 1}.$$

Overall, the correlation is

$$\frac{1 + \delta_x}{2^r} - \frac{1 - \frac{1+\delta_x}{2^r}}{2^r - 1} = \frac{\delta_x}{2^r - 1}.$$

This completes the proof of Lemma 2. ∎

We use Lemma 2 to complete the proof of the Many Hidden Bits Technical Theorem. The rest of the proof is a direct application of the Hidden Bit Technical Theorem (page 65). S^A works as follows. Run the oracle adversary described in the Hidden Bit Technical Theorem, making oracle queries to M^A and with input parameter set to $\delta/(2^r - 1)$, to create a list L. By the analysis given in Hidden Bit Technical Theorem, L has the property that if $\delta'_x \geq \delta/(2^r - 1)$ then $x \in L$ with probability at least $1/2$. Lemma 2 implies that if $\delta_x \geq \delta$ then $\delta'_x \geq \delta_x/(2^r - 1)$. It follows that, with probability at least $1/2$, $x \in L$. This completes the proof of the Many Hidden Bits Technical Theorem. ∎

Lecture 8

Overview

We introduce notions of statistical and computational entropy. We introduce universal hash functions and show how entropy can be smoothed using hashing.

Statistical Entropy

Definition (Shannon entropy): Let $\mathcal{D}_n$ be a distribution on $\{0,1\}^n$ and let $X \in_{\mathcal{D}_n} \{0,1\}^n$. For all $x \in \{0,1\}^n$, the *information* of x with respect to X is defined as

$$\mathrm{infor}_X(x) = \log(1/\Pr_X[X = x]) = -\log(\Pr_X[X = x]).$$

We can view $\mathrm{infor}_X(X)$ as a random variable defined in terms of X. The *entropy* of X is defined as the expected information of X, i.e.,

$$\mathrm{ent}(X) = \mathrm{E}_X[\mathrm{infor}_X(X)] = \sum_{x \in \{0,1\}^n} \Pr_X[X = x] \cdot \mathrm{infor}_X(x).$$

We use $\mathrm{ent}(\mathcal{D}_n)$ and $\mathrm{ent}(X)$ interchangeably. ♣

Note that if $\Pr[X = x] = 0$ then $\mathrm{infor}_X(x) = \infty$. The correct default in this case is to let $\Pr[X = x] \cdot \mathrm{infor}_X(x) = 0$.

Example : If $X \in_{\mathcal{U}} \{0,1\}^n$ then $\mathrm{ent}(X) = n$. More generally, if S is any finite set and $X \in_{\mathcal{U}} S$ then $\mathrm{ent}(X) = \log(\sharp S)$. In particular, if X always takes on a particular value with probability 1, then $\mathrm{ent}(X) = 0$.

Example : Let X be the random variable defined as follows.

$$X = \begin{cases} 0^n & \text{with probability } 1/2 \\ \langle 1, x \rangle, x \in \{0,1\}^{n-1} & \text{with probability } 1/2^n \end{cases}$$

Then, $\mathrm{ent}(X) = (n+1)/2$.

———∞———

We use the following fact to derive inequalities.

Fact : For all $z > 0$, $\ln(z) \leq z - 1$.

Exercise 31 : Prove the above fact. ♠

Lemma : If X is a random variable on $\{0,1\}^n$ then $\text{ent}(X) \le n$.

PROOF:

$$
\begin{aligned}
\text{ent}(X) - n &= \sum_{x \in \{0,1\}^n} \Pr_X[X = x] \cdot (\text{infor}_X(x) - n) \\
&= \log(e) \cdot \sum_{x \in \{0,1\}^n} \Pr_X[X = x] \cdot \ln(1/(\Pr_X[X = x]2^n)) \\
&\le \log(e) \cdot \sum_{x \in \{0,1\}^n} \Pr_X[X = x] \cdot (1/(\Pr_X[X = x]2^n) - 1) \\
&= \log(e) \cdot (1 - 1) = 0.
\end{aligned}
$$

The inequality follows from the fact stated above. ∎

This lemma shows that $X \in_{\mathcal{U}} \{0,1\}^n$ has the most entropy among all random variables distributed on $\{0,1\}^n$.

Exercise 32 : Let X and Y be independent random variables and let $Z = \langle X, Y \rangle$. Show that $\text{ent}(Z) = \text{ent}(X) + \text{ent}(Y)$. ♠

Definition (information divergence): Let X and Y be random variables distributed on $\{0,1\}^n$. The *information divergence* of Y with respect to X is defined as

$$
\sum_{x \in \{0,1\}^n} \Pr_X[X = x] \cdot \log(\Pr_X[X = x] / \Pr_Y[Y = x]).
$$

♣

Intuitively, the information divergence is small if the distribution Y places almost as much probability on each value as X does. For example, if $\Pr[Y = x] \ge \Pr[X = x]/2^\alpha$ for all $x \in \{0,1\}^n$ then the information divergence is at most α.

Kullback-Liebler information divergence inequality : For any pair of random variables X and Y distributed on $\{0,1\}^n$, the information divergence of Y with respect to X is greater than or equal to zero.

Exercise 33 : Prove the Kullback-Liebler information divergence inequality.

Hint : Use the fact stated above. ♠

Exercise 34 : Let X and Y be random variables that are not necessarily independent and let $Z = \langle X, Y \rangle$. Show that $\text{ent}(Z) \le \text{ent}(X) + \text{ent}(Y)$. ♠

Definition (prefix free encoding): A *prefix free encoding* of $\{0,1\}^n$ is a function f that maps $\{0,1\}^n$ to $\{0,1\}^*$ with the property that, for any $x, y \in \{0,1\}^n$, $x \neq y$, $f(x)$ is not a prefix of $f(y)$, i.e., there is no $z \in \{0,1\}^*$ such that $\langle f(x), z \rangle = f(y)$. ♣

Kraft inequality : For any prefix free encoding f of $\{0,1\}^n$,

$$\sum_{x \in \{0,1\}^n} 2^{-\|f(x)\|} \leq 1.$$

Exercise 35 : Prove the Kraft inequality. ♠

Let X be a random variable defined on $\{0,1\}^n$. The average length of a prefix free encoding f with respect to X is $E_X[\|f(X)\|]$. A good question to consider is, given a distribution on elements of a set, what encoding of the elements as strings is shortest on average with respect to the distribution? Exercise 36 shows that the Shannon entropy provides a lower bound on this quantity.

Exercise 36 : Prove that for all prefix free encodings f and for all random variables X,

$$E_X[\|f(X)\|] \geq \mathrm{ent}(X).$$

Hint : Use the Kraft inequality and the Kullback-Liebler information divergence inequality. ♠

Let $X_1, \ldots, X_n$ be independent identically distributed $\{0,1\}$-valued random variables, such that

$$p = \Pr_{X_1}[X_1 = 1] = \cdots = \Pr_{X_n}[X_n = 1].$$

Let

$$X = \sum_{i \in \{1, \ldots, n\}} X_i.$$

Let $1 \geq r > p$ and let Y be a $\{0,1\}$-valued random variable such that $\Pr_Y[Y = 1] = r$. The following lemma is a Chernoff type inequality.

Lemma : Let a be the information divergence of X_1 with respect to Y as defined above. Then, $\Pr_X[X \geq rn] \leq 2^{-an}$.

PROOF: For any $s > 1$,

$$\Pr_X[X \geq rn] = \Pr_X[s^X \geq s^{rn}] = \Pr_X[s^{X-rn} \geq 1].$$

From Markov's inequality (page 10), it follows that

$$\Pr_X[s^{X-rn} \geq 1] \leq E_X[s^{X-rn}]$$

$$= \prod_{i \in \{1,\dots,n\}} E_{X_i}[s^{X_i-r}] = (ps^{1-r} + (1-p)s^{-r})^n.$$

Let

$$s = \frac{r}{1-r} \cdot \frac{1-p}{p}.$$

Note that, because $r > p$, $s > 1$. For this value of s, the base of the rightmost expression can be simplified to

$$b = \left(\frac{r}{p}\right)^{-r} \cdot \left(\frac{1-p}{1-r}\right)^{1-r}.$$

The claim follows because $b = 2^{-a}$.   ■

Corollary : If $p = 1/2$ and $r > 1/2$ then $\Pr_X[X \geq rn] \leq 2^{n(\mathrm{ent}(Y)-1)}$.

Computational Entropy

One way of viewing a pseudorandom generator mapping n bits to $\ell(n)$ bits is that it accepts a distribution with entropy n (the uniform distribution on $\{0,1\}^n$) and stretches it to a distribution that looks like a distribution with entropy $\ell(n) > n$ (the uniform distribution on $\{0,1\}^{\ell(n)}$). Intuitively, a pseudoentropy generator is similar. It accepts a distribution with entropy n and stretches it to a distribution that looks like a distribution with entropy $\ell(n) > n$; but the difference is that it is not necessarily the uniform distribution.

Definition (computational entropy): Let $f : \{0,1\}^n \to \{0,1\}^{\ell(n)}$ be a **P**-time function ensemble with security parameter n. Let $X \in_{\mathcal{U}} \{0,1\}^n$ and $\mathcal{D}_n = f(X)$. We say f has **S**(n)-*secure computational entropy* $p(n)$ if there is a **P**-samplable probability ensemble $\mathcal{E}_n$ such that:

- $\mathrm{ent}(\mathcal{E}_n) \geq p(n)$.

- $\mathcal{D}_n$ and $\mathcal{E}_n$ are **S**(n)-secure computationally indistinguishable.

We say the computational entropy of f is non-uniform if $\mathcal{E}_n$ is not necessarily **P**-samplable.   ♣

Definition (pseudoentropy generator): Let $f : \{0,1\}^n \to \{0,1\}^{\ell(n)}$ be a **P**-time function ensemble with security parameter n. Let $p(n)$ be a non-negligible parameter. We say that f is a $\mathbf{S}(n)$-*secure* $p(n)$-*pseudoentropy generator* if f has $\mathbf{S}(n)$-secure computational entropy $n + p(n)$. ♣

In this definition, $p(n)$ is meant to measure the amount by which the entropy expands through the application of f, i.e., the input to f contains n bits of private entropy, and $f(X)$ looks like it has $n + p(n)$ bits of entropy, and thus the seeming expansion in entropy is $p(n)$.

A pseudorandom generator is a special case of a pseudoentropy generator where $\mathcal{E}_n$ is the uniform distribution on $\{0,1\}^{\ell(n)}$ for some $\ell(n) > n$ and $\ell(n) = n^{\mathcal{O}(1)}$. In this case, $p(n) = \ell(n) - n$.

Construction of pseudoentropy generator from a one-way one-to-one function : Let $f : \{0,1\}^n \to \{0,1\}^{\ell(n)}$ be a one-way one-to-one function with security parameter n. Let $x \in \{0,1\}^n$ and $z \in \{0,1\}^n$. Define **P**-time function ensemble $g(x,z) = \langle f(x), x \odot z, z \rangle$, where x is a private input and z is a public input, and thus the security parameter is $\| x \| = n$.

Theorem 8.1 : If f is a one-way one-to-one function then g is a pseudoentropy generator. The reduction is poly-preserving.

PROOF: Let $X \in_{\mathcal{U}} \{0,1\}^n$, $Z \in_{\mathcal{U}} \{0,1\}^n$ and $B \in_{\mathcal{U}} \{0,1\}$. Let $\mathcal{D}_n = g(X,Z)$ and let $\mathcal{E}_n = \langle f(X), B, Z \rangle$. The Hidden Bit Corollary (page 73) shows that if f is a one-way function then $\mathcal{D}_n$ and $\mathcal{E}_n$ are computationally indistinguishable. If, in addition, f is a one-to-one function, then since $\text{ent}(f(X)) = n$, it is easy to see that $\text{ent}(f(X), B, Z) = 2n + 1$. Thus, g has computational entropy $2n + 1$. On the other hand, the input entropy to g is only $2n$ bits, and thus g is a 1-pseudoentropy generator. The strength of the reduction from f to g follows from the Hidden Bit Corollary. ■

Alternative Notions of Entropy and Universal Hashing

In many of our constructions, a key idea is to apply a hash function to a random variable to extract its entropy in a usable form. For this purpose, it is useful to consider the following alternative notions of entropy.

Definition (minimum entropy): Define the *minimum entropy* of X as $\text{ent}_{\min}(X) = \min\{\text{infor}_X(x) : x \in \{0,1\}^n\}$, where X is a random variable defined on $\{0,1\}^n$. ♣

Definition (Renyi entropy): Let X and Y be independent and identically distributed random variables. Define the *Renyi entropy* of X as

$$\text{ent}_{\text{Ren}}(X) = -\log(\Pr_{X,Y}[X = Y]).$$

♣

Exercise 37 : Prove that for any random variable X,

$$\text{ent}_{\text{Ren}}(X)/2 \leq \text{ent}_{\min}(X) \leq \text{ent}_{\text{Ren}}(X) \leq \text{ent}(X).$$

♠

This shows that $\text{ent}_{\min}(X)$ and $\text{ent}_{\text{Ren}}(X)$ are the same within a factor of two. However, in general these two quantities can be substantially smaller than $\text{ent}(X)$, e.g., in the second example given in this lecture, $\text{ent}(X) = (n+1)/2$ whereas $\text{ent}_{\min}(X) = 1$.

Definition (universal hash function): Let $h : \{0,1\}^{\ell(n)} \times \{0,1\}^n \rightarrow \{0,1\}^{m(n)}$ be a **P**-time function ensemble. For fixed $y \in \{0,1\}^{\ell(n)}$, we view $h(y, x)$ as a function $h_y(x)$ of x that maps (or hashes) n bits to $m(n)$ bits. Let $Y \in_{\mathcal{U}} \{0,1\}^{\ell(n)}$. We say h is a (pairwise independent) *universal hash function* if, for all $x \in \{0,1\}^n$, $x' \in \{0,1\}^n \setminus \{x\}$, for all $a, a' \in \{0,1\}^{m(n)}$,

$$\Pr_Y[(h_Y(x) = a) \wedge (h_Y(x') = a')] = 1/2^{2m(n)},$$

i.e., h_Y maps every distinct pair x and x' independently and uniformly. ♣

Typically, the description y of the hash function is a public string, whereas the input x to the hash function is private.

Pairwise independent construction of random variables : Let $h : \{0,1\}^{\ell(n)} \times \{0,1\}^n \rightarrow \{0,1\}^{m(n)}$ be a universal hash function. A sequence of pairwise independent random variables can be constructed as follows. For all $i \in \{0,1\}^n$, define the value of the ith random variable X_i at sample point $y \in \{0,1\}^{\ell(n)}$ as $X_i(y) = h_y(i)$. Let $Y \in_{\mathcal{U}} \{0,1\}^{\ell(n)}$. By the properties of universal hash functions, it can be verified that the set of random variables $\{X_i(Y) : i \in \{0,1\}^n\}$ are pairwise independent and uniformly distributed in $\{0,1\}^{m(n)}$.

——————∞——————

The following two constructions of universal hash functions have several nice properties.

Inner product construction of a hash function : Given that we want to hash n bit strings to $m(n)$ bit strings, we define **P**-time function ensemble $h : \{0,1\}^{\ell(n)} \times \{0,1\}^n \to \{0,1\}^{m(n)}$ as follows, where $\ell(n) = (n+1) \cdot m(n)$. Let $x \in \{0,1\}^n$ and $y \in \{0,1\}^{(n+1) \times m(n)}$. Define

$$h_y(x) = \langle x, 1 \rangle \odot y.$$

(An alternative way to do this would be to let $y \in \{0,1\}^{n \times m(n)}$ and define $h_y(x) = x \odot y$. The only flaw with this is that $x = 0^n$ is mapped to $0^{m(n)}$ independently of y.) Note the similarity between this construction, the Inner Product Space (page 58) and the Generalized Inner Product Space (page 66).

Linear polynomial construction of a hash function : Given that we want to hash n bit strings to $m(n)$ bit strings, we define **P**-time function ensemble $h : \{0,1\}^{2\ell(n)} \times \{0,1\}^n \to \{0,1\}^{m(n)}$ as follows, where $\ell(n) = \max\{m(n), n\}$. In Lecture 5 we give a representation of the field elements of $\mathrm{GF}[2^{\ell(n)}]$ as $\{0,1\}^{\ell(n)}$ and describe an efficient way to compute field operations given this representation (page 58). Let $y = \langle y_1, y_2 \rangle \in \{0,1\}^{2 \times \ell(n)}$. Consider the polynomial

$$p_y(\zeta) = y_1 \zeta + y_2$$

over $\mathrm{GF}[2^{\ell(n)}]$, where y_1 and y_2 are considered as elements of $\mathrm{GF}[2^{\ell(n)}]$. Given $x \in \{0,1\}^n$, $h_y(x)$ is evaluated by considering x as an element of $\mathrm{GF}[2^{\ell(n)}]$ and computing $p_y(x)$, then $p_y(x)$ is interpreted as an $\ell(n)$-bit string and $h_y(x)$ is set to $p_y(x)_{\{1,\dots,m(n)\}}$. The advantage of this scheme over the first is that it takes fewer bits to describe the hash function. Note the similarity between this construction and the Linear Polynomial Space (page 57).

---∞---

Hereafter, whenever we refer to universal hash functions, we mean one of the constructions given above. However, any universal hash functions that satisfy the required properties may be used. When we introduce a universal hash function, the length of the description of the hash function is implicitly determined by the input and output lengths of the hash function. For example, if we say $h : \{0,1\}^{\ell(n)} \times \{0,1\}^n \to \{0,1\}^{m(n)}$ is a universal hash function, then $\ell(n)$ is implicitly defined in terms of n and $m(n)$ as described above.

The Smoothing Entropy Theorem given below is central to the development in the next few lectures. This theorem can be interpreted as follows. Suppose we have a random variable X with Renyi entropy at least m, but this entropy is in unusable form because the distribution on X is far from uniform. Let $h : \{0,1\}^\ell \times \{0,1\}^n \to \{0,1\}^{m-2e}$ be a universal hash function, where $\ell = \ell(n)$, $m = m(n)$ are functions of n, and

$e = e(n)$ is a small positive integer that controls the tradeoff between the uniformity of the output bits and the amount of entropy lost in the smoothing process. Let $Y \in_{\mathcal{U}} \{0,1\}^{\ell}$. The Smoothing Entropy Theorem shows that $h_Y(X)$ is essentially uniformly distributed independently of Y. Thus, we have managed to convert almost all the Renyi entropy of X into uniform random bits while maintaining our original supply of uniform random bits used to choose Y.

Smoothing Entropy Theorem : Let m be a positive integer and let X be a random variable defined on $\{0,1\}^n$ such that $\mathrm{ent}_{\mathrm{Ren}}(X) \geq m$. Let $e > 0$ be a positive integer parameter. Let $h : \{0,1\}^{\ell} \times \{0,1\}^n \rightarrow \{0,1\}^{m-2e}$ be a universal hash function. Let $Y \in_{\mathcal{U}} \{0,1\}^{\ell}$ and let $Z \in_{\mathcal{U}} \{0,1\}^{m-2e}$. Then, $\langle h_Y(X), Y \rangle$ and $\langle Z, Y \rangle$ are at most $2^{-(e+1)}$ statistically distinguishable.

PROOF: Let $s = m - 2e$. For all $y \in \{0,1\}^{\ell}$, $a \in \{0,1\}^s$ and $x \in \{0,1\}^n$, define $\chi(h_y(x) = a) = 1$ if $h_y(x) = a$ and 0 otherwise. We want to show that

$$1/2 \cdot \mathrm{E}_Y \left[\sum_{a \in \{0,1\}^s} \left| \mathrm{E}_X[\chi(h_Y(X) = a)] - 2^{-s} \right| \right] \leq 2^{-(e+1)}.$$

We show below that for all $a \in \{0,1\}^s$,

$$\mathrm{E}_Y[|\mathrm{E}_X[\chi(h_Y(X) = a)] - 2^{-s}|] \leq 2^{-(s+e)},$$

and from this the proof follows.

For any random variable Z, $\mathrm{E}_Z[|Z|^2] \geq \mathrm{E}_Z[|Z|]^2$ from Jensen's Inequality (page 12). From this it follows that $\mathrm{E}_Z[|Z|] \leq \mathrm{E}_Z[Z^2]^{1/2}$. Letting $Z = \mathrm{E}_X[\chi(h_Y(X) = a)] - 2^{-s}$, we see that it is sufficient to show for all $a \in \{0,1\}^s$,

$$\mathrm{E}_Y[(\mathrm{E}_X[\chi(h_Y(X) = a)] - 2^{-s})^2] \leq 2^{-2(s+e)}.$$

Let $X' \in_{\mathcal{D}} \{0,1\}^n$. Using some elementary expansion of terms, and rearrangements of summation, we can rewrite the lefthand side as

$$\mathrm{E}_{X,X'}[\mathrm{E}_Y[(\chi(h_Y(X) = a) - 2^{-s})(\chi(h_Y(X') = a) - 2^{-s})]].$$

For each fixed value of X to x and X' to x', where $x \neq x'$, the expectation with respect to Y is zero because of the pairwise independence property of universal hash functions. For each fixed value of X to x and X' to x', where $x = x'$,

$$\mathrm{E}_Y[(\chi(h_Y(x) = a) - 2^{-s})^2] = 2^{-s}(1 - 2^{-s}) \leq 2^{-s}.$$

Because $\text{ent}_{\text{Ren}}(X) \geq m$, it follows that $\Pr_{X,X'}[X = X'] \leq 2^{-m}$. Thus, the entire sum is at most $2^{-(m+s)}$ which is equal to $2^{-2(s+e)}$ by the definition of s. ∎

To be able to use the Smoothing Entropy Theorem, it is $\text{ent}_{\text{Ren}}(X)$ that must be large. In situations where $\text{ent}(X)$ is large and $\text{ent}_{\text{Ren}}(X)$ is small, we apply the universal hash function to the random variable Y defined as the concatenation of $k(n) = n^{\mathcal{O}(1)}$ independent copies of X to extract approximately $k(n)\text{ent}(X)$ uniformly distributed bits from Y. The Shannon to Renyi Theorem (page 101) essentially shows that $\text{ent}_{\text{Ren}}(Y) \approx k(n) \cdot \text{ent}(X)$, and thus the Smoothing Entropy Theorem applies.

Lecture 9

Overview

We describe two reductions from a one-way one-to-one function to a pseudorandom generator: The first is a weak-preserving reduction and the second is a poly-preserving reduction. We describe a poly-preserving reduction from a one-way regular function to a pseudorandom generator.

A pseudorandom generator from a one-way one-to-one function

We give two reductions from a one-way one-to-one function f to a pseudorandom generator g. The first reduction is only weak-preserving, whereas the second is poly-preserving. The first reduction is an immediate application of the Smoothing Entropy Theorem (page 86).

Definition (diagonal entries of a matrix): If $a \in \{0,1\}^{n \times n}$ then $\mathrm{diag}(a) = \langle a_{1,1}, a_{2,2}, \ldots, a_{n,n} \rangle$. ♣

weak-preserving construction for one-to-one functions : Let $f : \{0,1\}^n \to \{0,1\}^{\ell(n)}$ be a one-way one-to-one function. Let $y \in \{0,1\}^{2n \times n}$ and define

$$f'(y) = \langle f(y_1), \ldots, f(y_{2n}) \rangle.$$

Let $h' : \{0,1\}^{\ell'(n)} \times \{0,1\}^{2n\ell(n)} \to \{0,1\}^{2n^2 - n}$ be a universal hash function. Let $z \in \{0,1\}^{n \times 2n}$ and define pseudorandom generator

$$g(y, z, y') = \langle h'_{y'}(f'(y)), \mathrm{diag}(y \odot z), z, y' \rangle,$$

where y is a private input and z and y' are public inputs, and thus the security parameter is $\|y\| = 2n^2$.

Theorem 9.1 : If f is a one-way one-to-one function then g is a pseudorandom generator. The reduction is weak-preserving.

PROOF: Let $X \in_{\mathcal{U}} \{0,1\}^n$, $Y \in_{\mathcal{U}} \{0,1\}^{2n \times n}$, $W \in_{\mathcal{U}} \{0,1\}^n$, $Z \in_{\mathcal{U}} \{0,1\}^{n \times 2n}$, $Y' \in_{\mathcal{U}} \{0,1\}^{\ell'(n)}$, $B \in_{\mathcal{U}} \{0,1\}$ and $C \in_{\mathcal{U}} \{0,1\}^{2n}$. Let $\mathcal{D}_n = \langle f(X), X \odot W, W \rangle$ and let $\mathcal{E}_n = \langle f(X), B, W \rangle$. The Hidden Bit Theorem (page 65) implies that $\mathcal{D}_n$ and $\mathcal{E}_n$ are computationally indistinguishable. Let

$$\mathcal{D}'_n = \langle f'(Y), \mathrm{diag}(Y \odot Z), Z \rangle$$

and let
$$\mathcal{E}'_n = \langle f'(Y), C, Z \rangle.$$

Exercise 28 (page 72) shows that $\mathcal{D}'_n$ and $\mathcal{E}'_n$ are computationally indistinguishable. (This is where there is a huge loss in security and why the reduction is only weak-preserving.) Because f is a one-to-one function,

$$\text{ent}(f'(Y)) = \text{ent}_{\text{Ren}}(f'(Y)) = \text{ent}_{\text{min}}(f'(Y)) = 2n^2.$$

Let $D \in_\mathcal{U} \{0,1\}^{2n^2-n}$. By the Smoothing Entropy Theorem (page 86),

$$\langle h'_{Y'}(f'(Y)), Y' \rangle$$

and
$$\langle D, Y' \rangle$$

are at most $2^{-n/2}$-statistically distinguishable. By Exercise 29 (page 72), and because $\mathcal{D}'_n$ and $\mathcal{E}'_n$ are computationally indistinguishable, the distribution

$$g(Y, Z, Y') = \langle h'_{Y'}(f'(Y)), \text{diag}(Y \odot Z), Z, Y' \rangle$$

is computationally indistinguishable from

$$\langle h'_{Y'}(f'(Y)), C, Z, Y' \rangle.$$

Since $\langle D, C, Z, Y' \rangle$ and this last distribution are at most $2^{-n/2}$-statistically distinguishable,
$$\langle D, C, Z, Y' \rangle$$

and
$$g(Y, Z, Y')$$

are computationally indistinguishable. But, $\langle D, C, Z, Y' \rangle$ is the uniform distribution, and g stretches the input by n bits. It follows that g is a pseudorandom generator. ∎

The reason the reduction is only weak-preserving is the typical reason, i.e., g breaks up its private information of length $N = 2n^2$ into many small pieces of length n and applies f to each small piece. The reduction has the property that if A is an adversary that breaks g with private information of length $N = 2n^2$, then S^A is an adversary that breaks f with private information of length only a polynomial fraction of this length, i.e., on inputs of length $\sqrt{N/2} = n$.

poly-preserving construction for one-to-one functions : Let $f : \{0,1\}^n \to \{0,1\}^{\ell(n)}$ be a one-way one-to-one function. Let $x \in \{0,1\}^n$. Let $r(n)$ be a positive integer-valued function. Let $z \in \{0,1\}^{n \times (r(n)+1)}$.

Let $h' : \{0,1\}^{\ell'(n)} \times \{0,1\}^{\ell(n)} \to \{0,1\}^{n-r(n)}$ be a universal hash function. Define pseudorandom generator

$$g(x, z, y') = \langle h'_{y'}(f(x)), x \odot z, z, y' \rangle,$$

where x is a private input and both z and y' are public inputs, and thus the security parameter is $\|x\| = n$. Note that g stretches the input by one bit.

Theorem 9.2 :   If f is a one-way one-to-one function then g is a pseudorandom generator. The reduction is poly-preserving when $r(n)$ is set to the logarithm of the security of f.

PROOF: Suppose f is $\mathbf{S}(n)$-secure and let $r(n) = \alpha \log(S(n))$ for some constant $0 < \alpha < 1$. Let $Y' \in_{\mathcal{U}} \{0,1\}^{\ell'(n)}$, $X \in_{\mathcal{U}} \{0,1\}^n$, $Y \in_{\mathcal{U}} \{0,1\}^{n-r(n)}$, $Z \in_{\mathcal{U}} \{0,1\}^{n \times (r(n)+1)}$ and $B \in_{\mathcal{U}} \{0,1\}^{r(n)+1}$. The Many Hidden Bits Theorem (page 75) shows that

$$\mathcal{D}_n = \langle f(X), X \odot Z, Z \rangle$$

and

$$\mathcal{E}_n = \langle f(X), B, Z \rangle$$

are $\mathbf{S}(n)^{\Omega(1)}$-secure computationally indistinguishable. Exercise 29 shows that

$$\mathcal{D}'_n = \langle h'_{Y'}(f(X)), X \odot Z, Z, Y' \rangle$$

and

$$\mathcal{E}'_n = \langle h'_{Y'}(f(X)), B, Z, Y' \rangle$$

are $(\mathbf{S}(n)^{\Omega(1)}/n^{\mathcal{O}(1)})$-secure computationally indistinguishable. Because f is one-to-one, $\mathrm{ent}_{\mathrm{Ren}}(f(X)) = n$, and consequently the Smoothing Entropy Theorem shows that $\mathcal{E}'_n$ and

$$\mathcal{E}''_n = \langle Y, B, Z, Y' \rangle$$

are at most $2^{-r(n)/2}$-statistically distinguishable. From this, and Exercises 26 (page 71) and 27 (page 71), $\mathcal{D}'_n$ and $\mathcal{E}''_n$ are $(\mathbf{S}(n)^{\Omega(1)}/n^{\mathcal{O}(1)})$-secure computationally indistinguishable. The proof follows because $\mathcal{D}'_n$ is the distribution on the output of g and the public input, and $\mathcal{E}''_n$ is the uniform distribution. ∎

Theorem 9.2 describes a poly-preserving reduction, but it has the slightly disturbing property that this is only the case if the security of f is known, i.e., $r(n)$ is computed from $\mathbf{S}(n)$, where f is a $\mathbf{S}(n)$-secure one-way function. This is a property not shared by previous reductions. If $r(n)$ is chosen too large relative to the security of f then g may not be at all secure, and also g is at most $(2^{\Omega(r(n))}/n^{\mathcal{O}(1)})$-secure. Although technically this property is not desirable, it is not as bad as it might seem at first glance. Whenever a one-way function is used to construct a pseudorandom generator, some assumption about the security of the one-way function needs to be made to be able to confidently use the pseudorandom generator for most practical applications. Therefore, for any reduction, there is typically an implicit requirement that some assumption about the security of the one-way function be made in advance.

Regular Functions

A function is regular if each element in the range of the function has the same number of preimages. We first describe some properties of regular function ensembles, and then in the next section we describe a poly-preserving reduction from any one-way regular function to a pseudorandom generator.

Definition (range and preimages of a function): Let $f : \{0,1\}^n \to \{0,1\}^{\ell(n)}$ be a function ensemble. Define

$$\text{range}_f(n) = \{f(x) : x \in \{0,1\}^n\}.$$

For each $y \in \text{range}_f(n)$, define

$$\text{pre}_f(y) = \{x \in \{0,1\}^n : f(x) = y\}.$$

♣

Definition (regular function ensemble): We say function ensemble $f : \{0,1\}^n \to \{0,1\}^{\ell(n)}$ is $\sigma(n)$-*regular* if $\sharp\text{pre}_f(y) = \sigma(n)$ for all $y \in \text{range}_f(n)$. ♣

An important quantity to consider is how much information is lost when f is applied to an input x. For functions that are one-to-one, an application of f results in no information loss, i.e., x is uniquely determined by $f(x)$. For functions that are not necessarily one-to-one, x is not uniquely determined by $f(x)$, and thus there is a loss of information. To quantify this notion, we introduce the degeneracy of a function.

Definition (degeneracy): Let $\mathcal{D}_n$ be a probability ensemble with output length n and let $X \in_{\mathcal{D}_n} \{0,1\}^n$. Define the *degeneracy* of f with respect to $\mathcal{D}_n$ as

$$\text{degen}_f^{\mathcal{D}_n}(n) = \text{E}_X[\text{infor}_X(X) - \text{infor}_{f(X)}(f(X))].$$

Thus, $\text{degen}_f^{\mathcal{D}_n}(n) = \text{ent}(X) - \text{ent}(f(X))$. When $\mathcal{D}_n$ is the uniform distribution on $\{0,1\}^n$, we write $\text{degen}_f(n)$ in place of $\text{degen}_f^{\mathcal{D}_n}(n)$. ♣

Exercise 38 : Let $X \in_\mathcal{U} \{0,1\}^n$. Show that

$$\mathrm{infor}_X(x) - \mathrm{infor}_{f(X)}(f(x)) = \log(\sharp\mathrm{pre}_f(f(x))).$$

This implies that if f is a $\sigma(n)$-regular function ensemble then $\mathrm{degen}_f(n) = \log(\sigma(n))$. ♠

The following exercise shows that the Hidden Bit Corollary (page 73) is trivial if $\mathrm{degen}_f(n)$ is large.

Exercise 39 : Let $f(x)$ be a $\sigma(n)$-regular function ensemble. Let $X \in_\mathcal{U} \{0,1\}^n$, $Z \in_\mathcal{U} \{0,1\}^n$ and $B \in_\mathcal{U} \{0,1\}$. Let $\mathcal{D}_n = \langle f(X), X \odot Z, Z \rangle$ and let $\mathcal{E}_n = \langle f(X), B, Z \rangle$. Show that $\mathcal{D}_n$ and $\mathcal{E}_n$ are at most $(1/\sqrt{\sigma(n)})$-statistically distinguishable. ♠

A pseudorandom generator from a one-way regular function

Let $f : \{0,1\}^n \to \{0,1\}^{\ell(n)}$ be a one-way regular function with degeneracy $\mathrm{degen}_f(n)$ and let $X \in_\mathcal{U} \{0,1\}^n$. The main differences overall between the one-to-one construction and the regular construction are the following: (1) Instead of hashing $n - r(n)$ bits out of $f(X)$, hash $n - \mathrm{degen}_f(n) - r(n)$ bits out of $f(X)$, i.e., $\mathrm{degen}_f(n)$ fewer bits than before. (2) Compensate for this loss in entropy by hashing $\mathrm{degen}_f(n) - r(n)$ bits out of X, and similar to before also produce an $2r(n)+1$ inner product bits of X. Using the Smoothing Entropy Theorem (page 86) and the Many Hidden Bits Theorem (page 75) we show that the resulting g is a pseudorandom generator.

poly-preserving construction for regular function ensembles : Let $f : \{0,1\}^n \to \{0,1\}^{\ell(n)}$ be a one-way $\sigma(n)$-regular function ensemble and let

$$d(n) = \lceil \log(\sigma(n)) \rceil = \lceil \mathrm{degen}_f(n) \rceil.$$

Let $r(n)$ be a positive integer-valued function and let $z \in \{0,1\}^{n \times (2r(n)+1)}$. Let $h' : \{0,1\}^{\ell'(n)} \times \{0,1\}^{\ell(n)} \to \{0,1\}^{n-d(n)-r(n)}$ be a universal hash function. Let $h'' : \{0,1\}^{\ell''(n)} \times \{0,1\}^n \to \{0,1\}^{d(n)-r(n)}$ be a universal hash function. Define pseudorandom generator

$$g(x, z, y', y'') = \langle h'_{y'}(f(x)), h''_{y''}(x), x \odot z, z, y', y'' \rangle,$$

where x is a private input and all other inputs are public, and thus the security parameter is $\|x\| = n$. Note that g stretches the input by one bit.

Theorem 9.3 : If f is a one-way regular function then g is a pseudo-random generator. The reduction is poly-preserving when $r(n)$ is set to the logarithm of the security of f.

PROOF: Suppose f is $\mathbf{S}(n)$-secure and let $r(n) = \alpha \log(S(n))$ for some constant $0 < \alpha < 1$. Let $X \in_{\mathcal{U}} \{0,1\}^n$, $Y' \in_{\mathcal{U}} \{0,1\}^{\ell'(n)}$, $B' \in_{\mathcal{U}} \{0,1\}^{n-d(n)-r(n)}$, $Y'' \in_{\mathcal{U}} \{0,1\}^{\ell''(n)}$, $B'' \in_{\mathcal{U}} \{0,1\}^{d(n)-r(n)}$, $Z \in_{\mathcal{U}} \{0,1\}^{n \times (2r(n)+1)}$ and $B \in_{\mathcal{U}} \{0,1\}^{2r(n)+1}$. Let $Y \in_{\mathcal{U}} \mathrm{range}_f(n)$. Then, because f is a regular function ensemble, $Y = f(X)$. For each $y \in \mathrm{range}_f(n)$, let $P(y) \in_{\mathcal{U}} \mathrm{pre}_f(y)$. Then, $\langle Y, P(Y) \rangle = \langle f(X), X \rangle$. Fix $y \in \mathrm{range}_f(n)$. Because $\sharp\mathrm{pre}_f(y) = \sigma(n)$, $\mathrm{ent}_{\mathrm{Ren}}(P(y)) = \log(\sigma(n))$. From the Smoothing Entropy Theorem (page 86),

$$\langle y, h''_{Y''}(P(y)), Y'' \rangle \text{ and } \langle y, B'', Y'' \rangle$$

are at most $2^{-r(n)/2}$-statistically distinguishable. From this,

$$\langle f(X), h''_{Y''}(X), Y'' \rangle = \langle Y, h''_{Y''}(P(Y)), Y'' \rangle$$

and

$$\langle f(X), B'', Y'' \rangle = \langle Y, B'', Y'' \rangle$$

are at most $2^{-r(n)/2}$-statistically distinguishable. It is easy to see that since $f(X)$ is one-way then so is $\langle f(X), B'', Y'' \rangle$. From this, and based on the value of $r(n)$, $\langle f(X), h''_{Y''}(X), Y'' \rangle$ is a $(\mathbf{S}(n)^{\Omega(1)}/n^{\mathcal{O}(1)})$-secure one-way function. From this, and the Many Hidden Bits Theorem (page 75),

$$\mathcal{D}_n = \langle f(X), h''_{Y''}(X), X \odot Z, Z, Y'' \rangle$$

and

$$\mathcal{E}_n = \langle f(X), h''_{Y''}(X), B, Z, Y'' \rangle$$

are $(\mathbf{S}(n)^{\Omega(1)}/n^{\mathcal{O}(1)})$-secure computationally indistinguishable. Using the Smoothing Entropy Theorem again, $\mathcal{E}_n$ and

$$\mathcal{E}'_n = \langle f(X), B'', B, Z, Y'' \rangle$$

are at most $2^{-r(n)/2}$-statistically distinguishable, and thus by Exercise 26 (page 71) and Exercise 27 (page 71), $\mathcal{D}_n$ and $\mathcal{E}'_n$ are $(\mathbf{S}(n)^{\Omega(1)}/n^{\mathcal{O}(1)})$-secure computationally indistinguishable. Using the Smoothing Entropy Theorem again,

$$\langle h'_{Y'}(f(X)), B'', B, Z, Y', Y'' \rangle$$

and

$$\mathcal{E}''_n = \langle B', B'', B, Z, Y', Y'' \rangle$$

are at most $2^{-r(n)/2}$-statistically distinguishable. Thus, since $\mathcal{D}_n$ and $\mathcal{E}'_n$ are computationally indistinguishable, using Exercise 29 (page 72), and then Exercises 26 and 27 again,

$$\mathcal{D}'_n = \langle h'_{Y'}(f(X)), h''_{Y''}(X), X \odot Z, Z, Y', Y'' \rangle$$

and $\mathcal{E}''_n$ are $(\mathbf{S}(n)^{\Omega(1)}/n^{\mathcal{O}(1)})$-secure computationally indistinguishable. The theorem follows because $\mathcal{D}'_n$ is the distribution on the output of g, and $\mathcal{E}''_n$ is the uniform distribution. ∎

Lecture 10

Overview

We define a false entropy generator, show how to construct a false entropy generator from any one-way function in the non-uniform sense, and show how to construct a pseudorandom generator from a false entropy generator. Together, this yields a non-uniform reduction from any one-way function to a pseudorandom generator.

Preliminary Discussion

As shown in Theorem 9.2 (page 90), there is a poly-preserving reduction from a one-way one-to-one function to a pseudorandom generator. The intuitive idea of the reduction from any one-way function to a pseudorandom generator is to construct an almost one-to-one one-way function from any one-way function and then apply result already discussed. The final construction yields a weak-preserving reduction from any one-way function to a pseudorandom generator.

Definition (rank of a preimage): Let $f : \{0,1\}^n \to \{0,1\}^{\ell(n)}$ be a function ensemble. For all $x \in \{0,1\}^n$, let

$$\mathrm{rank}_f(x) = \#\{x' \in \mathrm{pre}_f(f(x)) : x' < x\}.$$

♣

Exercise 40 : Suppose $f : \{0,1\}^n \to \{0,1\}^{\ell(n)}$ is a one-way function and $\mathrm{rank}_f(x)$ is a **P**-time function ensemble. Prove that $g(x) = \langle f(x), \mathrm{rank}_f(x) \rangle$ is a one-way one-to-one function.

Hint : Suppose $f(x)$ is a $\sigma(n)$-regular function ensemble. Let A be an adversary for g as a one-way function. Consider the oracle adversary S that works as follows. On input y, S^A chooses randomly $\alpha \in_{\mathcal{U}} \{1, \ldots, \sigma(n)\}$ and queries A with input $\langle y, \alpha \rangle$. Let $X \in_{\mathcal{U}} \{0,1\}^n$. Show that the success probability of S^A for inverting $f(X)$ is the same as the success probability of A for inverting $g(X)$. The more general case, when f is not a regular function ensemble, is based on this idea. ♠

Based on Exercise 40 and Theorem 9.2 (page 90), it is easy to see that if f is a one-way function and $\mathrm{rank}_f(x)$ is a **P**-time function ensemble then there is a poly-preserving reduction from f to a pseudorandom generator. The problem is that $\mathrm{rank}_f(x)$ in general is not a **P**-time function ensemble. However, as the following theorem shows, we can simulate appending $\text{rank}_f(x)$ to $f(x)$ assuming that the following is a P-time function ensemble.

Definition $(d(f(x)))$:   For all $x \in \{0,1\}^n$, define

$$d(f(x)) = \lceil \log(\sharp\text{pre}_f(f(x))) \rceil .$$

♣

Construction when d is a P-time function ensemble :   Let $f : \{0,1\}^n \to \{0,1\}^{\ell(n)}$ be a one-way function and suppose $d(f(x))$ is a P-time function ensemble. Let $r(n)$ be a positive integer parameter. (There is a tradeoff in the reduction which depends on $r(n)$ between how close g is to a one-to-one function and how much security is lost in the reduction.) Let $h : \{0,1\}^{\ell'(n)} \times \{0,1\}^n \to \{0,1\}^{n+r(n)}$ be a universal hash function (page 84). Let $y' \in \{0,1\}^{\ell'(n)}$. Define one-way function

$$g(x,y') = \langle f(x), h_{y'}(x)_{\{1,\ldots,d(f(x))+r(n)\}}, y' \rangle,$$

where x is a private input and y' is public, and thus the security parameter is $\|x\| = n$.

Theorem 10.1 :   If f is a one-way function then g is a one-way function that is almost one-to-one, i.e., $\text{degen}_g(n) \leq 2^{-r(n)+2}$. The reduction is poly-preserving, with a loss of security that is proportional to $2^{r(n)}$.

PROOF:

$(g$ **is one-way) :**   Let $X \in_{\mathcal{U}} \{0,1\}^n$ and $Y' \in_{\mathcal{U}} \{0,1\}^{\ell'(n)}$. Suppose there is an adversary A that inverts $g(X,Y')$ with probability $\delta(n)$. We describe an oracle adversary S such that S^A is an adversary for f. The input to S^A is $f(x)$, where $x \in_{\mathcal{U}} \{0,1\}^n$.

Adversary S^A on input $f(x)$: .

Choose $y' \in_{\mathcal{U}} \{0,1\}^{\ell'(n)}$.

Choose $d \in_{\mathcal{U}} \{0,\ldots,n\}$.

Choose $b \in_{\mathcal{U}} \{0,1\}^{d+r(n)}$

Let $x' = A(f(x), b, y')$.

If $f(x') = f(x)$ then output x'.

Consider a fixed $y \in \mathrm{range}_f(n)$. Let $P(y) \in_{\mathcal{U}} \mathrm{pre}_f(y)$. Thus,

$$\mathrm{ent}_{\mathrm{Ren}}(P(y)) = \log(\sharp\mathrm{pre}_f(y)) \geq d(y) - 1.$$

Fix $s(n) = \lceil \log(1/\delta(n)) \rceil$ and let

$$
\begin{aligned}
a_0(y) &= d(y) - 1 - 2s(n), \\
a_1 &= 2s(n) + r(n) + 1, \\
a(y) &= a_0(y) + a_1 = d(y) + r(n).
\end{aligned}
$$

Let $B_0(y) \in_{\mathcal{U}} \{0,1\}^{a_0(y)}$, $B_1 \in_{\mathcal{U}} \{0,1\}^{a_1}$ and $B(y) = \langle B_0(y), B_1 \rangle$. By the Smoothing Entropy Theorem (page 86) it follows that

$$\langle y, h_{Y'}(P(y))_{\{1,\dots,a_0(y)\}}, Y' \rangle \text{ and } \langle y, B_0(y), Y' \rangle$$

are at most $2^{-s(n)-1}$-statistically distinguishable. By choice of $s(n)$, $2^{-s(n)-1} \leq \delta(n)/2$. On the other hand, for a fixed value of Y' to y' and for a fixed value of $P(y)$ to z, with probability 2^{-a_1} a randomly chosen value of B_1 for b_1 has the property that it is equal to $h_{y'}(z)_{\{a_0(y)+1,\dots,a(y)\}}$. Let Y be distributed according to $f(X)$. By definition, A inverts on input $\langle Y, h_{Y'}(P(Y))_{\{1,\dots,d(Y)+r(n)\}}, Y' \rangle$ with probability $\delta(n)$. With probability $1/(n+1)$, d is chosen to be $d(Y)$, in which case the input generated by S^A is $\langle Y, B(Y), Y' \rangle$. The probability that A inverts on this input is at least $(\delta(n) - 2^{-s(n)-1})2^{-a_1}$. The first term in the product is because A can only behave differently with respect to the first bits set to $\langle y, h_{Y'}(P(y))_{\{1,\dots,a_0(y)\}}, Y' \rangle$ or to $\langle y, B_0(y), Y' \rangle$ by at most the statistical distance between these two distributions, which is at most $2^{-s(n)-1}$. The second term is the probability that, when B_1 is set to b_1, b_1 is equal to the correct a_1-bit extension. Because $\delta(n) - 2^{-s(n)-1} \geq \delta(n)/2$, overall, S^A inverts $f(x)$ for randomly chosen $x \in_{\mathcal{U}} \{0,1\}^n$ with probability at least

$$\frac{\delta(n)^3}{(n+1)2^{r(n)+2}}.$$

The claim follows because the running time of S^A is essentially the same as the running time of A.

(**g is almost one-to-one**) : Let $X \in_{\mathcal{U}} \{0,1\}^n$, $X' \in_{\mathcal{U}} \{0,1\}^n$, $Y' \in_{\mathcal{U}} \{0,1\}^{\ell'(n)}$ and $Y'' \in_{\mathcal{U}} \{0,1\}^{\ell'(n)}$. Then, $\mathrm{ent}_{\mathrm{Ren}}(g(X,Y')) = -\log(\alpha)$ where

$$\alpha = \Pr_{X,Y',X',Y''}[g(X,Y') = g(X',Y'')].$$

It is not hard to verify that $\alpha = 2^{-\ell'(n)} \Pr_{X,X',Y'}[g(X,Y') = g(X',Y')]$. The probability that $X = X'$ is $1/2^n$, in which case $g(X,Y') = g(X',Y')$. The only other way that $g(X,Y') = g(X',Y')$ can occur is if

$$X' \in \text{pre}_f(X) \setminus \{X\}$$

and

$$h_{Y'}(X)_{\{1,\ldots,d(f(X))+r(n)\}} = h_{Y'}(X')_{\{1,\ldots,d(f(X'))+r(n)\}}.$$

The probability that $X' \in \text{pre}_f(X) \setminus \{X\}$ is $(\sharp\text{pre}_f(X) - 1)/2^n$. For a fixed value of X to x and X' to $x' \in \text{pre}_f(x) \setminus \{x\}$, the probability that

$$h_{Y'}(x)_{\{1,\ldots,d(f(x))+r(n)\}} = h_{Y'}(x')_{\{1,\ldots,d(f(x'))+r(n)\}}$$

is $2^{-d(f(x))-r(n)}$ by the pairwise independent property of universal hash functions. Since

$$(\sharp\text{pre}_f(x) - 1)/2^{d(f(x))} \leq 1,$$

it follows that

$$\Pr_{X,X',Y'}[g(X,Y') = g(X',Y')] \leq 2^{-n}(1 + 2^{-r(n)}).$$

The proof follows because this implies

$$\text{ent}_{\text{Ren}}(g(X,Y')) \geq \ell'(n) + n - \log(1 + 2^{-r(n)}) \geq \ell'(n) + n - 2^{-r(n)+2},$$

whereas if g were a one-to-one function its Renyi entropy would be $\ell'(n) + n$. From this it follows that $\text{degen}_g(n) \leq 2^{-r(n)+2}$. ∎

Exercise 41 : Let $f : \{0,1\}^n \to \{0,1\}^{\ell(n)}$ be a one-way function and suppose $d(f(x))$ is a **P**-time function ensemble. Use Theorem 10.1 to show that there is a poly-preserving reduction from f to a pseudorandom generator. ♠

A false entropy generator

Intuitively, g is a false entropy generator if the distribution $g(X)$ is computationally indistinguishable from a distribution $\mathcal{E}_n$ such that the entropy of $\mathcal{E}_n$ is greater than the entropy of $g(X)$. (In contrast, the stricter requirement for a pseudoentropy generator is that the entropy of $\mathcal{E}_n$ is greater than the entropy of the input X to g.)

Definition (false entropy generator): Let $g : \{0,1\}^n \to \{0,1\}^{\ell(n)}$ be a **P**-time function ensemble and let $X \in_{\mathcal{U}} \{0,1\}^n$. Let $p(n)$ be a non-negligible parameter. We say g has $\mathbf{S}(n)$-*secure false entropy* $p(n)$

if g has $\mathbf{S}(n)$-secure computational entropy $\mathrm{ent}(g(X)) + p(n)$. The false entropy is non-uniform if the computational entropy is non-uniform. ♣

The main idea behind the construction of g given below is to guess the value of $d(f(x))$ randomly.

Construction of g from a one-way function f : Let $f : \{0,1\}^n \to \{0,1\}^{\ell(n)}$ be a one-way function. Let $r(n)$ be a positive integer. Let $t(n) = \lceil \log(n + r(n) + 1) \rceil$ and let $d \in \{0, \ldots, 2^{t(n)} - 1\}$. Let $z \in \{0,1\}^{n \times r(n)}$. Let $h : \{0,1\}^{\ell'(n)} \times \{0,1\}^n \to \{0,1\}^{2^{t(n)}-1}$ be a universal hash function (page 84). Define false entropy generator

$$g(x, d, z, y') = \langle f(x), h_{y'}(x)_{\{1,\ldots,d\}}, x \odot z, d, z, y' \rangle$$

where x is a private input and the rest of the inputs are public, and thus the security parameter is $\|x\| = n$.

Theorem 10.2 : If f is a one-way function then g is a false entropy generator. The reduction is weak-preserving and non-uniform.

PROOF: Let $X \in_{\mathcal{U}} \{0,1\}^n$, $D \in_{\mathcal{U}} \{0,1\}^{t(n)}$, $Z \in_{\mathcal{U}} \{0,1\}^{n \times r(n)}$. $Y' \in_{\mathcal{U}} \{0,1\}^{\ell'(n)}$ and $B \in_{\mathcal{U}} \{0,1\}^{r(n)}$. Let

$$\mathcal{D}_n = g(X, D, Z, Y').$$

Let $\mathcal{E}_n$ be the same as $\mathcal{D}_n$ except that if $D = d(f(x)) + r(n)$ then $X \odot Z$ is replaced with B. Let $\mathcal{D}'_n$ be $\mathcal{D}_n$ conditional on $D = d(f(x)) + r(n)$, i.e.,

$$\mathcal{D}'_n = \langle f(X), h_{Y'}(X)_{\{1,\ldots,d(f(X))+r(n)\}}, X \odot Z, D, Z, Y' \rangle.$$

Similarly, let $\mathcal{E}'_n$ be $\mathcal{E}_n$ conditional on

$$d = d(f(x)) + r(n),$$

i.e.,

$$\mathcal{E}'_n = \langle f(X), h_{Y'}(X)_{\{1,\ldots,d(f(X))+r(n)\}}, B, D, Z, Y' \rangle.$$

The first part of the Theorem 10.1 (page 96) combined with the Many Hidden Bits Technical Theorem (page 75) shows that $\mathcal{D}'_n$ and $\mathcal{E}'_n$ are computationally indistinguishable, and the second part of the Theorem 10.1 shows that the amount of Renyi entropy added by $X \odot Z$ to $\mathcal{D}'_n$ is $\mathcal{O}(2^{-r(n)})$, whereas the amount of Renyi entropy added by B to $\mathcal{E}'_n$ is $r(n)$. Thus, $\mathrm{ent}_{\mathrm{Ren}}(\mathcal{E}'_n) \geq \mathrm{ent}_{\mathrm{Ren}}(\mathcal{D}'_n) + (r(n) - 1)$. Since $\mathcal{E}_n$ and $\mathcal{D}_n$ are exactly the same distribution when $D \neq d(f(x)) + r(n)$, and since $\mathcal{D}'_n$ and $\mathcal{E}'_n$ are computationally indistinguishable, it follows that $\mathcal{D}_n$ and $\mathcal{E}_n$ are computationally indistinguishable. Furthermore, since

$D = d(f(x)) + r(n)$ with probability $2^{-t(n)}$, it follows that $\text{ent}_{\text{Ren}}(\mathcal{E}_n) \geq \text{ent}_{\text{Ren}}(\mathcal{D}_n) + 2^{-t(n)}(r(n) - 1)$. ∎

The g described above is a non-uniform false entropy generator. The non-uniformity is because $\mathcal{E}_n$ is not necessarily **P**-samplable because $d(f(x))$ is not necessarily a **P**-time function ensemble. There is a known weak-preserving uniform reduction from a one-way function to a false entropy generator, but for brevity this reduction is omitted from these lectures.

A pseudorandom generator from a false entropy generator

We give a weak-preserving reduction from a false entropy generator to a pseudorandom generator. The reduction is non-uniform if the false entropy generator is non-uniform, and it is uniform otherwise. Combining this with the reduction given in the previous section from a one-way function to a non-uniform false entropy generator yields a weak-preserving non-uniform reduction from a one-way function to a pseudorandom generator.

Construction of a pseudorandom generator g from a false entropy generator f : Let $f : \{0,1\}^n \rightarrow \{0,1\}^{\ell(n)}$ be a $p(n)$-false entropy generator. For simplicity, we assume that $p(n) \leq 1$. Let $X \in_{\mathcal{U}} \{0,1\}^n$ and let $\mathcal{D}_n = f(X)$. As in the definition of a false entropy generator, let $\mathcal{E}_n$ be the distribution that is computationally indistinguishable from $\mathcal{D}_n$ such that $\text{ent}(\mathcal{E}_n) \geq \text{ent}(f(X)) + p(n)$. Let $d(n) = \text{degen}_f(n)$ be the degeneracy of f on inputs of length n. Let

$$k(n) = (n/p(n))^6,$$

$$y \in \{0,1\}^{k(n) \times n}$$

and

$$f'(y) = \langle f(y_1), \ldots, f(y_{k(n)}) \rangle.$$

Let $h : \{0,1\}^{\ell'(n)} \times \{0,1\}^{\ell(n)k(n)} \rightarrow \{0,1\}^{(n-d(n)+p(n))k(n) - 2k(n)^{5/6}}$, and $h' : \{0,1\}^{\ell''(n)} \times \{0,1\}^{nk(n)} \rightarrow \{0,1\}^{d(n)k(n) - 2k(n)^{5/6}}$

be universal hash functions. Define pseudorandom generator

$$g(y, y', y'') = \langle h_{y'}(f'(y)), h'_{y''}(y), y', y'' \rangle,$$

where y is a private input and both y' and y'' are public inputs, and thus the security parameter is $\|y\| = nk(n)$. Note that g stretches the input by $p(n)k(n) - 4k(n)^{5/6}$, and this is at least 1 for $n > 4$. The proof that

g is a pseudorandom generator requires the following technical theorem, which will not be proven.

Shannon to Renyi Theorem : Let Z be a random variable on domain $\{0,1\}^{\ell(n)}$ such that $\text{ent}(Z) \leq n$. Then, for any $k(n) \geq n^6$ there is a random variable Y such that

- Y is a distribution on $\{0,1\}^{\ell(n)k(n)}$.

- $\text{ent}_{\min}(Y) \geq k(n) \cdot \text{ent}(Z) - k(n)^{5/6}$.

- Y and the product distribution of $k(n)$ independent copies of Z are at most $2^{-k(n)^{1/3}}$-statistically distinguishable.

$$\text{---------}\infty\text{---------}$$

Theorem 10.3 : If f is a false entropy generator then g is a pseudorandom generator. The reduction is a weak-preserving. If the false entropy of f is non-uniform then the reduction is non-uniform.

PROOF: The proof proceeds in three steps. Let

$$Y \in_{\mathcal{U}} \{0,1\}^{k(n) \times n},$$

$$Y' \in_{\mathcal{U}} \{0,1\}^{\ell'(n)},$$

$$R_1 \in_{\mathcal{U}} \{0,1\}^{n-d(n)+p(n))k(n)-2k(n)^{5/6}},$$

$$Y'' \in_{\mathcal{U}} \{0,1\}^{\ell''(n)}, \text{ and}$$

$$R_2 \in_{\mathcal{U}} \{0,1\}^{d(n)k(n)-2k(n)^{5/6}}.$$

Step 1 : We claim that $\langle f'(Y), h'_{Y''}(Y), Y'' \rangle$ is at most $2^{1-k(n)^{1/3}}$-statistically distinguishable from $\langle f'(Y), R_2, Y'' \rangle$. To see this, consider how much entropy remains in $\langle h'_{Y''}(Y), Y'' \rangle$ after $f'(Y)$ is seen. We start with the string $\langle Y, Y'' \rangle$, a total of $nk(n) + \ell''(n)$ bits of entropy. After seeing $f'(Y)$, we will have used up $k(n) \cdot \text{ent}(f(X)) = k(n)(n - d(n))$ of those bits. Therefore, by using the Smoothing Entropy Theorem (page 86), we should be able to squeeze out about $d(n)k(n)$ more bits of entropy out of the original string.

To be more precise, define $W = f'(Y)$. For a fixed value of W to w, let

$$\text{pre}_{f'}(w) = \{y \in \{0,1\}^{k(n) \times n} : f'(y) = w\}.$$

For each w, let

$$V(w) \in_{\mathcal{U}} \text{pre}_{f'}(w)$$

and let
$$R(w) = \text{ent}_{\text{Ren}}(V(w)).$$

Then,
$$R(w) = \sum_{i=1}^{k(n)} \log(\sharp\text{pre}_f(w_i)).$$

Furthermore, because
$$\mathbf{E}_X[\log(\sharp\text{pre}_f(f(X)))] = \text{degen}_f(n) = d(n),$$

it follows that
$$\mathbf{E}_W[R(W)] = k(n)d(n).$$

Moreover, $R(W)$ is the sum of $k(n)$ independent random variables, and the range of each random variable is between 0 and n. Thus, using Chernoff bounds it is possible to show that

$$\Pr_W[R(W) < k(n)d(n) - k(n)^{5/6}] \le 2^{-k(n)^{1/3}}.$$

By the Smoothing Entropy Theorem (page 86), for all w such that $R(w) \ge k(n)d(n) - k(n)^{5/6}$,

$$\langle w, h'_{Y''}(V(w)), Y'' \rangle$$

is at most $2^{-k(n)^{1/3}}$-statistically distinguishable from

$$\langle w, R_2, Y'' \rangle.$$

The claim of Step 1 follows.

Step 2 : Let $Z_1, \ldots, Z_{k(n)} \in_{\mathcal{E}_n} \{0,1\}^{\ell(n)}$ and let $Z = \langle Z_1, \ldots, Z_{k(n)} \rangle$. We claim that $f'(Y)$ is computationally indistinguishable from Z. This claim follows from Exercise 28 (page 72) when $\mathcal{E}_n$ is **P**-samplable. The claim is true only in a non-uniform sense when $\mathcal{E}_n$ is not **P**-samplable, i.e., as is the case for the construction of a false entropy generator from a one-way function described in Theorem 10.2. Thus, the entire reduction is non-uniform when $\mathcal{E}_n$ is not **P**-samplable. Whether or not $\mathcal{E}_n$ is **P**-samplable, this step is where the loss of security occurs to make the reduction only weak-preserving.

Step 3 : We claim that $\langle h_{Y'}(Z), Y' \rangle$ is at most $2^{1-k(n)^{1/3}}$-statistically distinguishable from $\langle R_1, Y' \rangle$. We would like to extract about $(n-d(n)+p(n))k(n)$ bits of entropy from Z by using a hash function. From the Shannon to Renyi Theorem, we obtain a probability distribution $\mathcal{E}'_n$ such that $\mathcal{E}'_n$ is at most $2^{-k(n)^{1/3}}$-statistically distinguishable from Z and such that a random variable Z' distributed according to $\mathcal{E}'_n$ has the property that

$$\mathrm{ent_{min}}(Z') \geq \mathrm{ent}(\mathcal{E}_n)k(n) - k(n)^{5/6}.$$

From the Smoothing Entropy Theorem (page 86), $\langle h_{Y'}(Z'), Y' \rangle$ is at most $2^{-k(n)^{1/3}}$-statistically distinguishable from $\langle R_1, Y' \rangle$. The claim of Step 3 follows.

We can put these three steps together to finish the proof as follows. Step 1 shows that

$$\langle f'(Y), h'_{Y''}(Y), Y', Y'' \rangle \text{ and}$$

$$\langle f'(Y), R_2, Y', Y'' \rangle$$

are statistically indistinguishable. Step 2 shows that

$$\langle f'(Y), R_2, Y', Y'' \rangle \text{ and}$$

$$\langle Z, R_2, Y', Y'' \rangle$$

are computationally indistinguishable. Thus, Exercise 27 shows that

$$\langle f'(Y), h'_{Y''}(Y), Y', Y'' \rangle \text{ and}$$

$$\langle Z, R_2, Y', Y'' \rangle$$

are computationally indistinguishable. Applying $h_{Y'}$ to the first $k(n)\ell(n)$ bits of these two distributions, Exercise 29 shows that

$$\langle h_{Y'}(f'(Y)), h'_{Y''}(Y), Y', Y'' \rangle \text{ and}$$

$$\langle h_{Y'}(Z), R_2, Y', Y'' \rangle$$

are computationally indistinguishable. Step 3 shows that

$$\langle h_{Y'}(Z), R_2, Y', Y'' \rangle \text{ and}$$

$$\langle R_1, R_2, Y', Y'' \rangle$$

are statistically indistinguishable. Using Exercise 27 again, it follows that

$$\langle h_{Y'}(f'(Y)), h'_{Y''}(Y), Y', Y'' \rangle \text{ and}$$

$$\langle R_1, R_2, Y', Y'' \rangle$$

are computationally indistinguishable, where the first probability ensemble is $g(Y, Y', Y'')$ and the second is a random string of the same length. From this it follows that g is a pseudorandom generator. ∎

Exercise 42 : Show that there is a linear-preserving reduction from a pseudorandom generator to a one-way function.   ♠

Research Problem 4 :   A good open question is whether or not there is a stronger reduction from a one-way function to a pseudorandom generator, i.e., one that is poly-preserving or even linear-preserving, as opposed to the weak-preserving reduction given above. One approach to this problem is to show a linear-preserving or poly-preserving reduction from an arbitrary one-way function to an almost one-to-one one-way function.   ♠

Lecture 11

Overview

We define a stream private key cryptosystem, define several notions of security, including passive attack and chosen plaintext attack, and design a stream private key cryptosystem that is secure against these attacks based on a pseudorandom generator.

Stream Private Key Cryptosystem

We now consider the basic scenario described in Lecture 1 that was our initial motivation for constructing a pseudorandom generator. Parties P_1 and P_2 initially establish a shared random private key x of length n using a private line, and then afterwards P_1 is able to send messages privately on a public line to P_2 of total length $p(n)$, where $p(n) > n$.

In a stream private key cryptosystem, the encryption of each message bit is a function of the private key x, the index of the message bit (indexed by the order in which the bits are sent over the public line) and the actual message bit itself. Similarly, the decryption of an encrypted message bit depends on the private key, the index and the encryption of the message bit. To properly encrypt and decrypt message bits using a stream system, both parties must keep track of the index of the bit they are sending/receiving at each point in time. The term *stream* is used because the encryption/decryption of the message bits depends on the index of the bit within the entire stream of bits.

Definition (stream private key cryptosystem): A *stream private key cryptosystem* is a protocol for a party P_1 to send message bits privately to P_2 that works as follows.

(initialization): P_1 and P_2 exchange information over a private line to establish a private key $x \in \{0,1\}^n$. Both P_1 and P_2 store x in their respective private memories, and $\|x\| = n$ is the security parameter.

(message sending): Let $E : \{0,1\}^n \times \{0,1\}^{\ell(n)} \times \{0,1\} \to \{0,1\}^{k(n)}$ and $D : \{0,1\}^n \times \{0,1\}^{\ell(n)} \times \{0,1\}^{k(n)} \to \{0,1\}$ be **P**-time function ensembles, where $\ell(n)$ is the logarithm of the total number of message bits to be sent. E and D have the property that, for all

$x \in \{0, 1\}^n$, for all $i \in \{0, 1\}^{\ell(n)}$ and for $b \in \{0, 1\}$,

$$D_x(i, E_x(i, b)) = b.$$

P_1 sends the encryption of the message bits in sequence to P_2 on a public line. When P_1 wants to send the ith bit m_i (where m_i is presumably stored in the private memory), P_1 computes $e_i = E_x(i, m_i)$ using its private computation device and sends e_i on a public line to P_2. Upon receiving e_i, and knowing that e_i is the ith encryption sent over the public line, P_2 can recover m_i by computing $D_x(i, e_i)$ using the private memory device, storing the result presumably in private memory.   ♣

We have said nothing so far about the security of a stream private key cryptosystem. In the remainder of the lecture, we introduce four notions of security; security against simple passive attack, passive attack, simple chosen plaintext attack, and chosen plaintext attack. It is not hard to see that simple passive attack is a special case of passive attack, that simple chosen plaintext attack is a special case of chosen plaintext attack, that simple passive attack is a special case of simple chosen plaintext attack, and that passive attack is a special case of chosen plaintext attack. Other implications are not so clear.

Simple Passive Attack

We first give a definition of what it means to be secure against a simple passive attack. This attack consists of randomly and privately choosing a bit b and letting the adversary A see the encryption of the message which consists of a string of bits all equal to b. The success of A is measured in terms of how well it predicts the bit b.

Definition (simple passive attack for a stream system): Let $A : \{0, 1\}^{p(n)k(n)} \to \{0, 1\}$ be function ensemble. The attack works as follows:

(choose a private key): Choose a private key $x \in_\mathcal{U} \{0, 1\}^n$.

(choose a private message): Let $m^0 = 0^{p(n)}$ and $m^1 = 1^{p(n)}$. Choose $b \in_\mathcal{U} \{0, 1\}$ privately, let $m = m^b$ be the private message, and let

$$e = \langle E_x(1, m_1), \ldots, E_x(p(n), m_{p(n)}) \rangle$$

be the encryption of m.

(predict the bit): The success probability of the adversary is

$$\delta(n) = |\mathrm{E}[\overline{A(e)} \cdot \overline{b}]|.$$

(See page 4 for a reminder of the $\{1, -1\}$-bit notation.)

The stream cryptosystem is $\mathbf{S}(n)$-*secure against simple passive attack* if every adversary has time-success ratio at least $\mathbf{S}(n)$. ♣

Intuitively, the adversary A is trying to guess whether the message is $m^0 = 0^{p(n)}$ or $m^1 = 1^{p(n)}$, and $\delta(n)$ measures the correlation between the answer of A and the true answer.

Construction of a Stream System

We describe a stream private key cryptosystem based on a pseudorandom generator. Ideally, the encryption of each message bit should itself be a single bit, and this is the case for the construction given here. (This is the same construction informally described in Lecture 1.)

Construction of a stream cryptosystem : Let $g : \{0,1\}^n \to \{0,1\}^{p(n)}$ be a pseudorandom generator, where $p(n)$ is the maximum number of bits to be sent using private key x. Let $m \in \{0,1\}^{p(n)}$ be the message. The encryption of m_i is $e_i = E_x(i, m_i) = g(x)_i \oplus m_i$. The decryption of e_i is $m_i = D_x(i, e_i) = g(x)_i \oplus e_i$.

———————∞———————

Intuitively, this cryptosystem is secure because each encrypted bit looks like a completely random bit independent of all other encrypted bits.

A typical way to implement this construction is to use a pseudorandom generator g that stretches its output by one bit and use this to iteratively produce the output bits of a pseudorandom generator that produces a large number of output bits, say $p(n)$. As each output bit of the generator is produced, it is immediately used to encrypt the next bit of the message. The description below is an adaptation of the Stretching Algorithm (page 54). The proof that it correctly implements the construction described above is based on Theorem 3.3 (page 43). We assume that the private key x has already been established between P_1 and P_2 and stored in both of their private memories. Let $m = \langle m_1, \ldots, m_{p(n)} \rangle$ be the private message that P_1 wants to send to P_2.

How P_1 encrypts and sends bits to P_2 :

For $i = 1, \ldots, p(n)$ do

    P_1 privately computes $y = g(x)$.

    P_1 privately computes $e_i = y_1 \oplus m_i$

       and sends e_i to P_2 on a public line.

    P_1 replaces x in private memory with $y_{\{2,\ldots,n+1\}}$.

On the receiving end, P_2 enacts a similar algorithm to decrypt encrypted message bits.

Theorem 11.1 :   If g is a pseudorandom generator then the stream private key cryptosystem described above is secure against simple passive attack. The reduction is linear-preserving.

PROOF: Suppose there is an adversary A that has success probability $\delta(n)$ with respect to a simple passive attack when g is used as desribed in the construction above, Let $X \in_{\mathcal{U}} \{0,1\}^n$, $B \in_{\mathcal{U}} \{0,1\}$. Let $B^{p(n)} \in \{0,1\}^{p(n)}$ be the string consisting of B repeated $p(n)$ times. Then,

$$\delta(n) = |\mathrm{E}_{X,B}\overline{[A(g(X) \oplus B^{p(n)}) \cdot \overline{B}]}| = 2 \cdot |\Pr_{X,B}[A(g(X) \oplus B^{p(n)}) = B] - 1/2|.$$

Without loss of generality, let

$$\delta(n)/2 = \Pr_{X,B}[A(g(X) \oplus B^{p(n)}) = B] - 1/2.$$

Let $Y \in_{\mathcal{U}} \{0,1\}^{p(n)}$. Because $Y \oplus B^{p(n)} \in_{\mathcal{U}} \{0,1\}^{p(n)}$ is distributed uniformly independent of B, and because $B \in_{\mathcal{U}} \{0,1\}$,

$$\Pr_{Y,B}[A(Y \oplus B^{p(n)}) = B] - 1/2 = 0.$$

Thus, the following adversary S^A can be used to distinguish $g(X)$ from Y. The input to S^A is z, where $z \in \{0,1\}^{p(n)}$.

Adversary S^A on input z :

Choose $b \in_{\mathcal{U}} \{0,1\}$.

If $A(z \oplus b^{p(n)}) = b$ then output 1 else output 0.

From the above calculations, S^A produces 1 with probability $1/2 + \delta(n)/2$ when the input is $g(X)$, and S^A produces 1 with probability $1/2$ when the input is Y. Thus, S^A has success probability $\delta(n)/2$ for distinguishing g as a pseudorandom generator. ∎

Passive Attack

We now give a more general definition of what it means to be secure against a passive attack. This attack is more general in the sense that the adversary is trying to predict a general bit-valued function b of a message generated privately by a function P based on a randomly chosen private input r. This definition is motivated by the following discussion.

No cryptosystem can protect the privacy of messages that P_1 wants to send to P_2 from an adversary unless the adversary has some a priori uncertainty about what the messages are. For example, if the adversary knows that P_1 is going to send the string of all zeroes then there is no uncertainty remaining in the message that can be hidden from A by any (even a perfect) encryption system.

We model this uncertainty by specifying a function ensemble P that produces a probability distribution on messages given a random input. We would be concerned about the security of the cryptosystem if the adversary is able to obtain any information, even a single bit of information, about the message that P_1 sends to P_2 based on its encryption. We model this by specifying a $\{0, 1\}$-valued function ensemble b that accepts as input the message, where the value of b is meant to be the bit of the message that the adversary is trying to obtain.

Informally, a passive attack works as follows. The overall adversary consists of three adversaries, P, b, and A. A private key x is chosen randomly and privately. The output of P defines a distribution on messages with respect to a random input. A message m is produced randomly and privately by P, and A receives the encryption $E_x(m)$. The adversary b accepts as input a message and produces a single bit, where $b(m)$ is the bit that A is interested in knowing about the message m. Based on $E_x(m)$, A produces $c \in \{0, 1\}$, which is meant to be a prediction of $b(m)$. The success probability is the covariance of c and $b(m)$ as defined below.

Definition (covariance): Let X and Y be $\{0, 1\}$-valued random variables that are not necessarily independent of one another and not necessarily uniformly distributed. The *covariance* of Y and X is defined as

$$\operatorname{covar}(X, Y) = E[Y(X - E[X])] = E[X \cdot Y] - E[X] \cdot E[Y].$$

Let Z be a random variable that is not necessarily independent of X and Y. The *conditional covariance* of X and Y given Z is defined as

$$\operatorname{covar}(X, Y | Z) = E_Z[E[X \cdot Y | Z] - E[X | Z] \cdot E[Y | Z]].$$

♣

Intuitively, $\operatorname{covar}(X, Y)$ is a measure of how well Y helps to predict the value of X. The covariance can be thought of as a generalization of the correlation defined on page 5) when X is not necessarily uniformly distributed. In the following examples, X and Y are $\{0, 1\}$-valued random variables.

- $\operatorname{covar}(X, Y) = 0$ iff X and Y are independent. This agrees with the intuition that if the random bits are independent then knowing the value of Y gives no additional information about the value of X.

- At the opposite end of the spectrum, if $Y \in_{\mathcal{U}} \{0, 1\}$ and $X = Y$, then $\operatorname{covar}(X, Y)$ is the maximum possible, i.e., $1/4$. Intuitively, this example should achieve the maximum possible covariance because a priori Y has the maximum amount of uncertainty possible for a single bit random variable, but knowing the value of Y completely determines the value of X.

- A little more generally, if $p = \Pr[Y = 1]$ and $X = Y$ then $\operatorname{covar}(X, Y) = p(1 - p)$. Note that the covariance goes to zero as p gets close to either zero or one. The intuitive reason is that as p goes to zero, the value of X is becomes more and more determined a priori, and thus knowing the value of Y cannot add much more information about the value of X.

Exercise 43 :   Given that $p = \Pr[Y = 1]$, prove that the maximum possible covariance $p(1 - p)$ is achieved when $Y = X$.   ♠

Intuitively, $\operatorname{covar}(X, Y | Z)$ is a measure of how well the value of Y helps to predict the value of X when the value of Z is already known. Here is an example that illustrate the difference between conditional covariance and covariance. Suppose $Z \in_{\mathcal{U}} \{0, 1\}^n$ and X and Y are both equal to the first bit Z_1 of Z. Then, $\operatorname{covar}(X, Y) = 1/4$ whereas $\operatorname{covar}(X, Y | Z) = 0$.

This agrees with the intuition that Y provides a lot of information about the value of X if nothing is known about X a priori, whereas Y provides no additional information about X if the value of Z is already known.

Definition (passive attack for a stream system):

Let $P : \{0,1\}^{s(n)} \rightarrow \{0,1\}^{p(n)}$, $b : \{0,1\}^{p(n)} \rightarrow \{0,1\}$, and $A : \{0,1\}^{p(n)k(n)} \rightarrow \{0,1\}$ be adversaries.

The attack works as follows:

(choose a private key): Choose a private key $x \in_{\mathcal{U}} \{0,1\}^n$.

(choose a private message): Choose $r \in_{\mathcal{U}} \{0,1\}^{s(n)}$ privately. Let $m = P(r)$ be the private message produced by P and let

$$e = \langle E_x(1, m_1), \ldots, E_x(p(n), m_{p(n)}) \rangle$$

be the encryption of m.

(predict the bit): The success probability of the adversary is

$$\delta(n) = |\text{covar}(A(e), b(m))|.$$

The run time $T(n)$ of the overall adversary includes the time to compute P, b and A. The stream cryptosystem is $\mathbf{S}(n)$-*secure against passive attack* if every adversary has time-success ratio at least $\mathbf{S}(n)$. ♣

The definition of security should imply the cryptosystem is not secure if the adversary is able to obtain x based on the attack, and this is the case: If the adversary has x then it can compute m (as described below) and output $b(m)$. Given that $\Pr[b(m) = 1] = p$, the success probability is $p(1-p)$, i.e., the maximum possible for the amount of uncertainty p in the value of $b(m)$. Here is how the adversary recovers m using x. For each $i \in \{1, \ldots, p(n)\}$, compute $e_i^0 = E_x(i, 0)$ and $e_i^1 = E_x(i, 1)$. Recall that the adversary also knows the encryption $e_i = E_x(i, m_i)$ of m_i from the attack. The adversary can deduce that m_i is equal to $j \in \{0,1\}$, where $e_i^j = e_i$.

Here is an example of a passive attack. P produces the uniform distribution on messages and b produces the first bit of the message, i.e., $b(m) = m_1$. In this case, the success of the attack is measured by how well A can predict the first bit of a uniformly chosen message based on its encryption.

Let $R \in_{\mathcal{U}} \{0,1\}^{s(n)}$ and let $p = \Pr[b(P(R)) = 1]$. Exercise 43 above shows that the success probability of any attack is at most $p(1-p)$. For an attack to have any chance of having a significant success probability, it is crucial that there is some uncertainty in the bit $b(P(R))$ that A is trying to predict. For example, if $P(R)$ puts probability one on the message $m = 0^{p(n)}$ then, no matter what b is, the success probability of the attack is zero.

Exercise 44 : Prove that if g is a pseudorandom generator then the stream private key cryptosystem described above is secure against passive attack. The reduction should be linear-preserving. ♠

Simple Chosen Plaintext Attack

In the stronger chosen plaintext attacks described below, the adversary is allowed to be quite invasive. As for the stream attack, we first give a simple definition before the more general definition. The overall adversary consists of three adversaries, M, P, and A. M is allowed to first interactively choose several messages and see their encryptions. Then, P specifies two possible messages to encrypt. One of the two message is chosen privately at random and the adversary A sees its encryption. Finally, A tries to predict which of the two messages was encrypted.

Definition (simple chosen plaintext attack for a stream system):

Let $M : \{0,1\}^{\log(p(n))} \times \{0,1\}^{s(n)} \times \{0,1\}^{p(n)k(n)} \to \{0,1\}$,
   $P : \{0,1\}^{s(n)} \times \{0,1\}^{p(n)k(n)} \to \{0,1\}^{2t(n)}$,
   $A : \{0,1\}^{s(n)} \times \{0,1\}^{p(n)k(n)} \times \{0,1\}^{t(n)k(n)} \to \{0,1\}$ be adversaries.

The attack works as follows.

(choose the private key): Choose a private key $x \in_{\mathcal{U}} \{0,1\}^n$.

(chosen plaintext attack): Choose $r \in_{\mathcal{U}} \{0,1\}^{s(n)}$. For $j = 1, \ldots, p(n)$, phase j works as follows. Let

$$e = \langle E_x(1, m_1), \ldots, E_x(j-1, m_{j-1}) \rangle$$

be the concatenation of the encryptions of the first $j-1$ message bits produced by M padded out with zeroes to a string of length $k(n)p(n)$. Then, $m_j = M(j, r, e)$. At the end of all $p(n)$ phases, let $m = \langle m_1, \ldots, m_{p(n)} \rangle$ be the message produced by M and let

$$e = \langle E_x(1, m_1), \ldots, E_x(p(n), m_{p(n)}) \rangle$$

be the encryption of m.

(choose the private message): Let $\langle m^0, m^1 \rangle = P(r, e)$ be two $t(n)$-bit messages produced by P. Choose $b \in_{\mathcal{U}} \{0, 1\}$ privately. Let $m' = m^b$ be the private message, and let

$$e' = \langle E_x(p(n) + 1, m_1'), \ldots, E_x(p(n) + t(n), m_{t(n)}') \rangle$$

be the encryption of m'.

(predict the bit): The success probability of the adversary is

$$\delta(n) = |\mathrm{E}[\overline{A(r, e, e')} \cdot \overline{b}]|.$$

The run time $T(n)$ of the overall adversary includes the time to compute M, P, b and A. The stream cryptosystem is $\mathbf{S}(n)$-*secure against chosen plaintext attack* if every adversary has time-success ratio at least $\mathbf{S}(n)$. ♣

From the definitions above, a simple passive attack is a special case of a simple chosen plaintext attack where M skips the chosen plaintext attack step. We could allow an even more invasive attack, where the adversary invokes a chosen plaintext attack both before and after seeing the encryption of the private message. The results given below all generalize to this case.

In the chosen plaintext attack step, M implicitly knows $\langle m_1, \ldots, m_{j-1} \rangle$, since this can be easily generated using M based on r and e. Similarly, in the step where the private message is chosen, P implicitly knows m, since this can be easily generated using M based on r and e. In the step where the adversary tries to predict the bit, A implicitly knows m, m^0 and m^1, since these can be easily generated using M and P based on r and e.

Exercise 45 : Prove that the previously described stream private key cryptosystem is secure against simple chosen plaintext attack. The reduction should be linear-preserving. ♠

Chosen Plaintext Attack

Using the same approach as used to go from simple passive attack to passive attack, we modify the definition of simple chosen plaintext attack to define chosen plaintext attack.

The overall adversary consists of four adversaries, M, P, b and A. Based on a random string r, M sequentially produces message bits and receives their encryptions: this is the chosen plaintext attack. At the end of this, P produces a message m' privately based on a new private random string

r', on the random string r used by M, and on the encryption e of the message m produced by M. It turns out that m can be easily generated from r and e, and thus implicitly m' also depends on m. Intuitively, P models the behavior of the party generating an important message after the chosen plaintext attack, and $b(m')$ is the bit that the adversary would like to predict. Let e' be the encryption of m'. A tries to predict $b(m')$ based on r, e and e'. Since m can be easily generated from r and e, A's prediction implicitly also depends on m.

Definition (chosen plaintext attack for a stream system):

Let $M : \{0,1\}^{\log(p(n))} \times \{0,1\}^{s(n)} \times \{0,1\}^{p(n)k(n)} \to \{0,1\}$,
$\quad P : \{0,1\}^{s(n)} \times \{0,1\}^{s(n)} \times \{0,1\}^{p(n)k(n)} \to \{0,1\}^{t(n)}$,
$\quad b : \{0,1\}^{t(n)} \to \{0,1\}$,
$\quad A : \{0,1\}^{s(n)} \times \{0,1\}^{p(n)k(n)} \times \{0,1\}^{t(n)k(n)} \to \{0,1\}$ be adversaries.

The attack works as follows.

(choose the private key): Choose a private key $x \in_{\mathcal{U}} \{0,1\}^n$.

(chosen plaintext attack): Choose $r \in_{\mathcal{U}} \{0,1\}^{s(n)}$. For $j = 1, \ldots, p(n)$, phase j works as follows. Let

$$e = \langle E_x(1, m_1), \ldots, E_x(j-1, m_{j-1}) \rangle$$

be the concatenation of the encryptions of the first $j-1$ message bits produced by M padded out with zeroes to a string of length $k(n)p(n)$. Then, $m_j = M(j, r, e)$. At the end of all $p(n)$ phases, let $m = \langle m_1, \ldots, m_{p(n)} \rangle$ be the message produced by M and let $e = \langle E_x(1, m_1), \ldots, E_x(p(n), m_{p(n)}) \rangle$ be the encryption of m.

(choose the private message): Choose $r' \in_{\mathcal{U}} \{0,1\}^{s(n)}$. Let $m' = P(r', r, e)$ be the private message produced by P, and let $e' = \langle E_x(p(n)+1, m_1'), \ldots, E_x(p(n)+t(n), m_{t(n)}') \rangle$ be the encryption of m'.

(predict the bit): The success probability of the adversary is

$$\delta(n) = |\mathrm{covar}(A(r, e, e'), b(m')|r, e)|.$$

The run time $T(n)$ of the overall adversary includes the time to compute M, P, b and A. The stream cryptosystem is $\mathbf{S}(n)$-*secure against chosen plaintext attack* if every adversary has time-success ratio at least $\mathbf{S}(n)$. ♣

Similar to the remarks made about simple attacks, a passive attack is a special case of a chosen plaintext attack where M skips the chosen plaintext attack step. Moreover, all the results given below can be generalized to the case where the adversary invokes a chosen plaintext attack both before and after seeing the encryption of the private message.

We now give three scenarios that fit into the above definition. In all scenarios, there is some bit α that the adversary would really like to know. For example, suppose a party P_1 is on the board of directors of a company and the adversary is a board member of a second rival company. Suppose the board of directors of the first company holds a private meeting to decide their strategy, e.g., to increase production of their existing product and flood the market to saturation at reduced prices ($\alpha = 0$) or to put into production a much improved version of their product at much higher prices ($\alpha = 1$). After the meeting, P_1 wants to send the outcome α to another board member P_2 who is away on an important business trip in a faraway city. Of course, the adversary is able to read all information on the public line used by P_1 to send to P_2. Suppose that a priori α is uniformly distributed from the point of view of the adversary. What the adversary wants to do at the end of the attack is to produce a bit c that is equal to α.

Example of passive attack : Suppose the adversary knows the first bit that P_1 sends to P_2 is α. Furthermore, the adversary knows the distribution on the remaining part of the message given the value of α, which is perhaps a summary of how the decision was made. The adversary has no interaction with P_1 other than to see the encrypted message sent to P_2. This is an example of a passive attack.

Example of chosen plaintext attack : In this example, the adversary is almost fully able to control what message P_1 sends to P_2, i.e., the adversary decides one by one (interactively, depending possibly upon encryptions of previous bits) on the values of all but one of the message bits and receives their encryptions. Finally, the last message bit is chosen privately at random, the adversary receives the encryption of this bit, and based on all this the adversary tries to predict its value. For example, suppose that although the adversary and P_1 are board members in rival companies, they are also good friends. The adversary knows that P_1 is quite a gossip, and before the board meeting the adversary has several conversations with P_1, and in each conversation the adversary tells P_1 a piece of gossip chosen by the adversary, knowing full well that P_1 immediately sends these conversations verbatim in encrypted form to P_2. Furthermore, the gossip that the adversary tells P_1 in a given conversation can depend on encryptions of pieces of gossip that P_1

previously sent to P_2. Finally, the board meeting is held and P_1 sends the encryption of the one bit outcome α to P_2 as the final bit of the entire message. Because the adversary is choosing the plaintext that is encrypted and sent over the line, this type of attack is an example of chosen plaintext attack.

Example where b is more complicated :   Suppose the adversary knows that P_1 is going to send α using the following strategy. P_1 chooses the first $\log(p(n))$ bits of the message uniformly at random, and then sends α in the bit position indexed by these $\log(p(n))$ bits. In this case, the $\{0, 1\}$-valued function that the adversary wants to predict is not a fixed message bit, but a more complicated function of the message bits.

Exercise 46 :   Prove that the stream private key cryptosystem described previously based on a pseudorandom generator is secure against chosen plaintext attack. The reduction should be linear-preserving.   ♠

When using a stream private key cryptosystem, each message bit that is encrypted is only private if it is stored in private memory. If an adversary manages to obtain the contents of part of the message, this does not automatically compromise the privacy of the rest of the message. On the other hand, the privacy of the entire message is lost if the adversary is able to obtain the private key x, and thus x must be kept private as long as any one of the many message bits encrypted using x is to be kept private. This explains why the security of the stream cryptosystem is parameterized in terms of $\| x \|$.

Lecture 12

Overview

We define a block cryptosystem and security against chosen plaintext attack. We show how to construct a pseudorandom function generator from a pseudorandom generator, and show how a pseudorandom function generator can be used to construct a block private key cryptosystem secure against chosen plaintext attack.

Block Private Key Cryptosystem

There are some practical problems with using a stream cryptosystem. First of all, because of the implicit indexing, both parties have to stay in lock step forever, and if transmissions get garbled at some point then they have to resynchronize somehow. One way to get around this problem is to send the index of the bit together with its encryption over the public line (implicitly the index is an output of the encryption function in any case), but this is a relatively clumsy and inefficient solution, e.g., the amount of information sent per encrypted message bit is rather large.

Perhaps a more serious problem is that a stream cryptosystem is inconvenient to use if two or more users want to encrypt and send messages using the same private key, even if they are willing to pay the price of sending an index with each encrypted message bit. The time to encrypt and decrypt using the stream system described on page 108 depends linearly on the difference between indices of successive message bits. To ensure a unique index for each message bit, large gaps between successive indices used by each party are typical, and encrypting using a stream system is not efficient in these circumstances.

These problems motivate the definition and construction of a block private key cryptosystem. The message is partitioned into blocks of equal length, and a unique index is associated with each message block. The index is used to encrypt the message block, and is sent in plaintext along with the encryption of the message block to allow decryption. Since an index is long, it is easy to ensure that all indices are unique. Using a block system, the effective rate of information transfer is reasonable. For example, if the index is the same length as the message block, and the encryption of a message block is the same length as the block itself, then the effective rate of communication is one actual bit received for every two bits sent. In practice, the length of the index is usually a small fraction of length of the message block. Recall that for the stream system described on page 108, the encryption time depended linearly on the value of the index, which in general is exponential in its length. In contrast, the time dependence on the index is linear in its length for all the block cryptosystems we construct.

Definition (block private key cryptosystem): A *block private key cryptosystem* for an ensemble of parties consists of the following.

(initialization): All parties exchange information over private lines to establish a private key $x \in \{0,1\}^n$. All parties store x in their respective private memories, and $\|x\| = n$ is the security parameter.

(message sending):

Let $E : \{0,1\}^n \times \{0,1\}^{\ell(n)} \times \{0,1\}^{q(n)} \rightarrow \{0,1\}^{k(n)}$ and
$D : \{0,1\}^n \times \{0,1\}^{\ell(n)} \times \{0,1\}^{k(n)} \rightarrow \{0,1\}^{q(n)}$

be **P**-time function ensembles. E and D have the property that, for all $x \in \{0,1\}^n$, for all $i \in \{0,1\}^{\ell(n)}$ and for all $m \in \{0,1\}^{q(n)}$,

$$D_x(i, E_x(i, m)) = m.$$

A party sends a message block $m \in \{0,1\}^{q(n)}$ by first choosing an index $i \in \{0,1\}^{\ell(n)}$ distinct from all other indices ever used with the same private key x and then privately computing $e = E_x(i, m)$ and sending $\langle i, e \rangle$ on a public line. Upon receiving $\langle i, e \rangle$, another party can recover m by computing $D_x(i, e)$ using the private memory device, storing the result presumably in private memory.   ♣

For all subsequent attacks we only describe a simple version of the attack where the adversary eventually generates one of two possible messages, one of the two is randomly and privately chosen, the adversary is given the encryption of the chosen message, and then eventually the adversary tries to predict which of the two messages was chosen and encrypted.

In all the attacks against block systems described below, the adversary is allowed to choose the indices for the message blocks, but with the restriction that the index specified for the privately chosen message must be distinct from all other indices. This restriction on the adversary is natural, because the index associated with a message is typically produced automatically by the party independent of the content of the message. Thus, even though the adversary may be able to exert considerable influence on the messages the party encrypts, not so much control is allowed for the index.

We now give two methods of implementing automatic indexing.

Unique id automatic indexing : A unique identifier u_j is associated with party P_j. P_j also uses an auxiliary variable v_j initialized to zero. When a message block is to be encrypted, P_j increments v_j by one and uses the index $\langle u_j, v_j \rangle \in \{0,1\}^{\ell(n)}$.

Randomized automatic indexing : When a message block is to be encrypted, P_j randomly chooses an index $i \in_{\mathcal{U}} \{0,1\}^{\ell(n)}$.

———————∞———————

Unique id automatic indexing guarantees that all indices are unique, even indices produced by different parties. An adversary attacking this scheme may have some control over indexing, e.g., by the choice of which party encrypts the message and by the order the messages are encrypted. However, the adversary cannot manage to have two messages encrypted using the same index, even two identical messages.

Randomized automatic indexing does not guarantee that all indices are unique. However, if $p(n)$ messages in total are encrypted using the same private key then the probability that some pair of messages have the same index is bounded by $p(n)^2/2^{\ell(n)}$. In all the arguments we give about the security of a system where we assume unique indexing, using randomized automatic indexing can increase the success probability of any adversary by at most $p(n)^2/2^{\ell(n)}$. For $\ell(n)$ sufficiently large, e.g., $\ell(n) = n$, the increase in success probability is exponentially small in n.

The following definition is a natural modification of the definition of security against simple chosen plaintext attack for stream cryptosystems given on page 112.

Definition (chosen plaintext attack for a block system):

Let $M : \{0,1\}^{\log(p(n))} \times \{0,1\}^{s(n)} \times \{0,1\}^{p(n)k(n)} \rightarrow \{0,1\}^{\ell(n)+q(n)}$,
$P : \{0,1\}^{s(n)} \times \{0,1\}^{p(n)k(n)} \rightarrow \{0,1\}^{\ell(n)+2q(n)}$,
$A : \{0,1\}^{s(n)} \times \{0,1\}^{p(n)k(n)} \times \{0,1\}^{k(n)} \rightarrow \{0,1\}$ be adversaries.

The attack works as follows.

(choose a private key): Choose a private key $x \in_{\mathcal{U}} \{0,1\}^n$.

(chosen plaintext attack): Choose $r \in_{\mathcal{U}} \{0,1\}^{s(n)}$. For $j = 1, \ldots, p(n)$, phase j works as follows. Let

$$e = \langle E_x(i_1, m_1), \ldots, E_x(i_{j-1}, m_{j-1}) \rangle$$

be the concatenation of the encryptions of the first $j - 1$ message blocks padded out with zeroes to a string of length $p(n)k(n)$. Then, $\langle i_j, m_j \rangle = M(j, r, e)$. At the end of all $p(n)$ phases, let $i = \langle i_1, \ldots, i_{p(n)} \rangle$, $m = \langle m_1, \ldots, m_{p(n)} \rangle$, and

$$e = \langle E_x(i_1, m_1), \ldots, E_x(i_{p(n)}, m_{p(n)}) \rangle.$$

(choose the private message block): Let $\langle i', m^0, m^1 \rangle = P(r, e)$ be the index and the pair of message blocks produced by P. It is required that P produce an index i' that is distinct from all of the $p(n)$ indices contained in i. Choose $b \in_{\mathcal{U}} \{0, 1\}$ privately, let $m' = m^b$ be the private message, and let $e' = E_x(i', m')$ be the encryption of m'.

(predict the bit): The success probability of the adversary is

$$\delta(n) = |\mathrm{E}[\overline{A(r, e, e')} \cdot \overline{b}]|.$$

The run time $T(n)$ of the overall adversary includes the time to compute M, P and A. The block cryptosystem is $\mathbf{S}(n)$-*secure against chosen plaintext attack* if every adversary has time-success ratio at least $\mathbf{S}(n)$. ♣

Note that when P generates the index i' and the two possible message blocks m^0 and m^1, P implicitly knows all indices i and message blocks m generated during the chosen plaintext attack, since these can be computed based on r and e using M. Similarly, when A is trying to predict b, A implicitly knows i, m, i', m^0 and m^1, since these can be computed based on r and e using M and P.

The definition of a chosen plaintext attack can be generalized in a natural way to allow the adversary to invoke a chosen plaintext attack both before and after seeing the encryption of the private message.

Interlude: A Story about Chosen Plaintext Attack

In a chosen plaintext attack the adversary is allowed complete control over what is encrypted during the attack. In practice, an adversary may be able to exert some control over which messages are encrypted, but may not be able to mount a full-fledged chosen plaintext attack against the cryptosystem. Thus, a weaker type of security than chosen plaintext attack may be sufficient in practice. However, a cryptosystem that is secure against chosen plaintext attack is also secure against these weaker types of attacks that may occur in practice.

An entertaining example of the feasibility of this kind of attack is from the movie "Midway". The situation is that both the Americans and the Japanese are using cryptosystems, and the Japanese have broken the American system, which the Americans know, but the Japanese don't know that the Americans know. On the other hand, the Americans have almost broken the Japanese cryptosystem, which the Japanese don't know. The "almost" is because although the Americans have cracked the basic Japanese cryptosystem, the Japanese use another level of encryption for names of places that the Americans haven't yet broken. At some point in time the Japanese send the following message using their cryptosystem: "We are going to attack Υ at the crack of dawn tomorrow." The Americans decrypt this message, except that they don't know the decryption of the place Υ. However, the Americans have a pretty good idea it is Midway Island that the Japanese are going to attack, so they send the following message over their cryptosystem, knowing that the Japanese will be able to decrypt the message: "There is a water shortage on Midway Island". Sure enough, the Japanese are able to decrypt the message, and they send out using their encryption system the message which the Americans are able to decrypt to: "There is a water shortage on Υ". This of course confirmed the American suspicions about where the attack was going to take place, and thus the Americans were able to concentrate their defense efforts on Midway Island, saving the day.

Of course this example is rather devious, and it is easy to think of other less nefarious means by which an interested bystander would be able to obtain an encryption of one or more messages of choice.

Exercise 47 : Watch the movie "Midway".   ♠

---

pseudorandom function generators

A block cryptosystem is perfectly secure if the encryption of each distinctly indexed block is random independent of all other encryptions. One immediate attempt to implement a block cryptosystem is to partition the output of a pseudorandom generator into equal length blocks and let the encoding of a block of the message be its exclusive-or with the corresponding block of the pseudorandom generator output. The problem with this approach is efficiency: a party may be forced to generate all preceding bits of the pseudorandom generator to be able to produce the block corresponding to the encryption, e.g., a party that receives an encryption block with index N might have to generate all $N - 1$ previous blocks of the output of the pseudorandom generator to be able to decrypt. This is especially a problem in a multiparty environment if several parties use the same pseudorandom generator (with the same private seed) to send encryptions among themselves, and each party uses a set of non-overlapping indices for the blocks.

A pseudorandom function generator overcomes these problems in a natural way, and can be used directly to efficiently implement a block cryptosystem. We construct a pseudorandom function generator based on a pseudorandom generator and then use a pseudorandom function generator to construct a secure block private key cryptosystem.

Definition (pseudorandom function generator): Let $f : \{0,1\}^n \times \{0,1\}^{\ell(n)} \rightarrow \{0,1\}^{k(n)}$ be a **P**-time function ensemble, where the first input is private and the second is public, and thus the security parameter is n. For fixed $x \in \{0,1\}^n$, we view $f(x,i)$ as a function $f_x(i)$ of i, and thus

$$f_x \in \mathbf{Fnc}{:}\{0,1\}^{\ell(n)} \rightarrow \{0,1\}^{k(n)}.$$

Let $X \in_{\mathcal{U}} \{0,1\}^n$ and

$$F \in_{\mathcal{U}} \mathbf{Fnc}{:}\{0,1\}^{\ell(n)} \rightarrow \{0,1\}^{k(n)}.$$

Let A be an oracle adversary that produces a single bit. The inputs and outputs of the oracle queries made by A are strings of length $\ell(n)$ and $k(n)$, respectively. The success probability of adversary A for f is

$$\delta(n) = |\Pr_X[A^{f_X}(n) = 1] - \Pr_F[A^F(n) = 1]|.$$

Then f is a $\mathbf{S}(n)$-*secure pseudorandom function generator* if every adversary A for f has time-success ratio at least $\mathbf{S}(n)$.   ♣

Intuitively, the first input x to f is a description of the function f_x that must always be kept secret for the security of the pseudorandom function generator to be maintained. The second input to f is an input to f_x, and this is considered public because the adversary is allowed to interactively specify many inputs to f_x and see the corresponding outputs during the attack.

A pseudorandom function generator is more powerful than a pseudorandom generator. One way of comparing a pseudorandom function generator with a pseudorandom generator is the following. The output of a pseudorandom generator g looks like a polynomial length random string to an adversary. In contrast, a pseudorandom function generator f looks like a random function when evaluated at a polynomial number of inputs interactively chosen by an adversary. The description of a truly random function is exponentially long, and thus in this sense f looks like an exponentially long random string to an adversary.

Construction of a pseudorandom function generator : Let g : $\{0,1\}^n \rightarrow \{0,1\}^{2n}$ be a pseudorandom generator that doubles the length of its input. When $x \in \{0,1\}^n$ then we write $g(x) = \langle g(x)_1, g(x)_2 \rangle \in \{0,1\}^{2 \times n}$. We use g to define a **P**-time function ensemble $f : \{0,1\}^n \times \{0,1\}^{\ell(n)} \rightarrow \{0,1\}^n$ with security parameter n as follows. Define $f_x \in$ **Fnc**:$\{0,1\}^{\le \ell(n)} \rightarrow \{0,1\}^n$ inductively on the length of the input as follows.

- $f_x(\lambda) = x$.

- Let $i \in \{0,1\}^{\le \ell(n)-1}$. The pair $(\langle i, 0 \rangle, \langle i, 1 \rangle)$ are the children of i, and i is the parent of $(\langle i, 0 \rangle, \langle i, 1 \rangle)$. Given that $f_x(i)$ has been defined, let

$$f_x(i, 0) = g(f_x(i))_1,$$

and

$$f_x(i, 1) = g(f_x(i))_2.$$

————————∞————————

Given $x \in \{0,1\}^n$ and $i \in \{0,1\}^{\ell(n)}$, the value of $f_x(i)$ can be computed by a party by making a total of $\ell(n)$ queries to g, i.e., for all $j = 1, \ldots, \ell(n)$, compute

$$f_x(i_{\{1,\ldots,j\}}) = g(f_x(i_{\{1,\ldots,j-1\}}))_{i_j}$$

as just described.

Theorem 12.1 : If g is a pseudorandom generator then f is a pseudorandom function generator. The reduction is poly-preserving.

PROOF: Let A be an adversary for f with success probability $\delta(n)$ and run time $T(n)$. Let $X \in_{\mathcal{U}} \{0,1\}^n$ and $F \in_{\mathcal{U}}$ **Fnc**:$\{0,1\}^{\ell(n)} \rightarrow \{0,1\}^n$. Let

$$q_0 = \Pr_X[A^{f_X}(n) = 1]$$

and

$$q_1 = \Pr_X[A^F(n) = 1].$$

and without loss of generality, $\delta(n) = q_0 - q_1$. Let $m(n)$ be the maximum over all functions $h \in$ **Fnc**:$\{0,1\}^{\ell(n)} \rightarrow \{0,1\}^n$ of the number of oracle queries A^h makes, and thus $m(n) \le T(n)$. The proof uses an argument that is similar in spirit to the proof of the Stretching Theorem (page 52). We describe an oracle adversary S such that S^A is an adversary for g. The input to S^A is $z \in \{0,1\}^{2 \times n}$. S^A will simulate A evaluating the oracle queries using a function $h \in$ **Fnc**:$\{0,1\}^{\le \ell(n)} \rightarrow \{0,1\}^n$ that is a hybrid between F and f_X. Initially, S^A sets the value of $h(\lambda)$ to randomly chosen $s \in_{\mathcal{U}} \{0,1\}^n$. Other input/output values of h are determined interactively as described below by S^A as the simulation of A proceeds. Let $i \in \{0,1\}^{\ell(n)}$ be an input to an oracle query that A makes. When S^A simulates computing the output of the oracle query, it computes in sequence the output value of h for the following $\ell(n)$ pairs of input values:

$$\langle\langle 0, 1\rangle, \langle\langle i_1, 0\rangle, \langle i_1, 1\rangle\rangle, \langle\langle i_{\{1,2\}}, 0\rangle, \langle i_{\{1,2\}}, 1\rangle\rangle,$$

$$\ldots, \langle\langle i_{\{1,\ldots,\ell(n)-1\}}, 0\rangle, \langle i_{\{1,\ldots,\ell(n)-1\}}, 1\rangle\rangle\rangle.$$

The output of the oracle query is $h(i)$. To make sure that S^A is always computing a function h, S^A stores all input/output pairs to h during the course of the simulation of A, and whenever S^A needs to compute an output to h for an input $u \in \{0,1\}^{\le \ell(n)}$, S^A first checks to see if u was a previous input to h, and if so it returns as the answer the previously stored output value.

During the course of the simulation, S^A occasionally has to decide on the output value of h for a new pair of input values $\langle\langle u, 0\rangle, \langle u, 1\rangle\rangle$ that have not been previously seen. The new pairs are ordered by occurrence within the simulation. When $\langle\langle u, 0\rangle, \langle u, 1\rangle\rangle$ occurs as a new pair during the simulation, then the value of $h(u)$ for the parent u has already been determined. For all $j = 1, \ldots, \ell(n)m(n)$, let $\langle\langle u_j, 0\rangle, \langle u_j, 1\rangle\rangle$ denote the jth new pair.

As stated above, S^A initially sets $h(\lambda) = s$, where $s \in_{\mathcal{U}} \{0,1\}^n$. The output value of h for new pairs is determined as follows by S^A. S^A randomly chooses

$$k \in_{\mathcal{U}} \{1, \ldots, \ell(n)m(n)\}$$

and

$$r \in_{\mathcal{U}} \{0,1\}^{2(k-1)\times n}.$$

For all $j = 1, \ldots, k-1$, when the jth new pair occurs during the simulation then S^A sets $h(u_j, 0) = r_{2j-1}$ and $h(u_j, 1) = r_{2j}$. When the kth new pair occurs during the simulation then S^A sets $h(u_k, 0) = z_1$ and $h(u_k, 1) = z_2$. (Recall that $z = \langle z_1, z_2\rangle$ is the string that A is trying to classify as either being truly random or as an output of g.) For all $j = k+1, \ldots, \ell(n)m(n)$, when the jth new pair occurs during the simulation then S^A sets $h(u_j, 0) = g(h(u_j))_1$ and $h(u_j, 1) = g(h(u_j))_2$.

The final output of S^A is the output of the simulated A.

For the analysis of S^A, let $X \in_{\mathcal{U}} \{0,1\}^n$, $Y \in_{\mathcal{U}} \{0,1\}^{2\times n}$ and let $F \in_{\mathcal{U}}$ **Fnc**$:\{0,1\}^{\le \ell(n)} \to \{0,1\}^n$. Let $p_{0,j}$ be the probability that the output of S^A is 1 when the distribution on the input z is $g(X)$ and when $k = j$. Let

$p_{1,j}$ be the probability that the output of S^A is 1 when the distribution on the input z is Y and when $k = j$. Then, the overall probability that the output of S^A is 1 when the distribution on the input z is $g(X)$ is

$$p_0 = \frac{1}{\ell(n)m(n)} \sum_{j=1}^{\ell(n)m(n)} p_{0,j},$$

whereas the overall probability that the output of S^A is 1 when the distribution on the input z is Y is

$$p_1 = \frac{1}{\ell(n)m(n)} \sum_{j=1}^{\ell(n)m(n)} p_{1,j}.$$

The key observations are:

- For all $j = 1, \ldots, \ell(n)m(n) - 1$, $p_{0,j+1} = p_{1,j}$. This is because:

  - When $k = j + 1$ and z distributed according to $g(X)$:

    * For the jth new pair: S^A chooses the value of h for these two inputs randomly.
    * For the $j + 1^{rst}$ new pair: S^A uses its input $g(X)$ to set the values of h for these two inputs.

  - When $k = j$ and z distributed according to Y:

    * For the jth new pair: S^A uses its input Y to set the values of h for these two inputs.
    * For the $j + 1^{rst}$ new pair: S^A uses the value of $g(h(u_j))$ to set the values of h for these two inputs, where $h(u_j)$ was set randomly previously (either to Y_1, Y_2 or else to a random value chosen by S^A.)

  Thus, after the $j+1^{rst}$ new pair, the distribution on values assigned to h for all subsequent new pairs is exactly the same in both cases, and each subsequent new pair follows exactly the same procedure to assign a value to h in both cases.

- $p_{0,1} = q_0$. This is because the simulation by S^A in this case is exactly the same as if though $h(\lambda) = X$ and for every subsequent new pair the value of h is computed by applying g to the parent of the pair, and thus the oracle queries are all computed according to f_X.

- $p_{1,\ell(n)m(n)} = q_1$. This is because the total number of oracle queries by A is at most $m(n)$, and since each such query causes at most $\ell(n)$ new pairs, the total number of new pairs is at most $\ell(n)m(n)$. Since S^A uses random values as the output of h for each new pair in this case, the oracle queries are being evaluated according to F.

Let $\delta'(n)$ be the success probability of S^A for distinguishing $g(X)$ and Y. Then,

$$\delta'(n) \;=\; p_0 - p_1 = \frac{1}{\ell(n)m(n)} \sum_{j=1}^{\ell(n)m(n)} p_{0,j} - p_{1,j}$$

$$= \; \frac{p_{0,1} - p_{1,\ell(n)m(n)}}{\ell(n)m(n)} = \frac{\delta(n)}{\ell(n)m(n)}.$$

The run time of S^A is $n^{\mathcal{O}(1)} \cdot T(n)$.   ∎

Construction of a Block Cryptosystem

Construction of a block cryptosystem from a pseudorandom function generator :   Let $f : \{0,1\}^n \times \{0,1\}^{\ell(n)} \to \{0,1\}^n$ be a pseudorandom function generator with security parameter $n.$. Define a block private key cryptosystem using f as follows. Let $x \in \{0,1\}^n$ be the private key. The encryption of message $m \in \{0,1\}^n$ with index $i \in \{0,1\}^{\ell(n)}$ is $e = m \oplus f_x(i)$. The decryption of e with index i is $e \oplus f_x(i)$.

Exercise 48 :   Prove that the block private key cryptosystem based on pseudorandom function generator f described above is secure against chosen plaintext attack. The reduction from the pseudorandom function generator f to the secure block private key cryptosystem should be linear-preserving.   ♠

Exercise 49 : Show there is a linear-preserving reduction from a pseudorandom function generator to a pseudorandom generator.   ♠

Exercise 50 :   Let $g(x)$ be a pseudorandom generator that doubles the length of its input, where if $x \in \{0,1\}^n$ then $g(x) \in \{0,1\}^{2\times n}$. Let $X \in_U \{0,1\}^n$. Describe how to use g to construct two sequences of random variables

$$Y_0(X), \ldots, Y_{2^n-1}(X) \in \{0,1\}^n$$

and
$$Z_0(X), \ldots, Z_{2^n-1}(X) \in \{0,1\}^n$$
with the following properties.

- *Easy to compute Y forward from Z:* Let $x \in \{0,1\}^n$. Given $i, j \in \{0,1\}^n$ with $i < j$, and given $Z_i(x)$ (but not x or anything else), $Y_j(x)$ is computable in $n^{\mathcal{O}(1)}$ time.

- *Hard to compute Y backward from Z:* Let A be an adversary that works as follows. Choose $x \in_{\mathcal{U}} \{0,1\}^n$ privately. A specifies $i, j \in \{0,1\}^n$ with $j < i$ and A receives $Z_i(x)$. The success probability of A is the probability that A is able to produce $Y_j(x)$. Describe an oracle adversary S such that if A has time-success ratio $R'(n)$ then S^A has time-success ratio $R(n)$ for g, where $R(n) = n^{\mathcal{O}(1)} \cdot \mathcal{O}(R'(n))$. ♠

Lecture 13

Overview

We define the notion of a pseudorandom invertible permutation generator and discuss applications to the construction of a block private key cryptosystem secure against chosen plaintext attack. We introduce a construction of a perfect random permutation based on a perfect random function.

pseudorandom invertible permutation generator

The Data Encryption Standard (DES) is a standard private key cryptosystem used in the United States by the business community. DES is the motivation for both the definition of a pseudorandom invertible permutation generator and the construction of a pseudorandom invertible permutation generator from a pseudorandom function generator. DES can be thought of as

$$g^{64} \subset \mathbf{Perm}{:}\{0,1\}^{64} \to \{0,1\}^{64},$$

where each function $g_x \in g^{64}$ in the family is indexed by a private key $x \in \{0,1\}^{56}$. The computation of g_x involves first expanding x to sixteen 48-bit strings using an easily computable rule, and this defines sixteen easily computable functions $f_1, \ldots, f_{16}$, each mapping $\{0,1\}^{32}$ to $\{0,1\}^{32}$. On input $y = \langle y_1, y_2 \rangle \in \{0,1\}^{2 \times 32}$ the value of $g_x(y)$ is computed as follows. Let $\ell_0 = y_1$ and $r_0 = y_2$ and, for all $i = 1, \ldots, 16$, $\ell_i = r_{i-1}$ and $r_i = \ell_{i-1} \oplus f_i(r_{i-1})$. Then, $g_x(y) = \langle \ell_{16}, r_{16} \rangle$.

One important property of DES is that if the private key x is known then it is easy to undo each of these steps with the inverse operation: $r_{i-1} = \ell_i$ and $\ell_{i-1} = r_i \oplus f_i(\ell_i)$. Thus, for each x, $g_x \in \mathbf{Perm}{:}\{0,1\}^{64} \to \{0,1\}^{64}$. Furthermore, the inverse permutation $\bar{g}_x$ of g_x is also easily computable given x. The way DES is used is as a block private key cryptosystem, with the property that encryption and decryption use the same private key and the length of the encryption of a block is exactly the same length as the block itself (this is the best that could be hoped for).

Whether or not DES is secure in a practical sense when used as a private key cryptosystem is debatable, but what is clear is that it has some desirable features. Some of these features motivate the definition of a pseudorandom invertible permutation generator and its construction from a pseudorandom function generator.

Definition (pseudorandom invertible permutation generator):
An invertible permutation generator $\langle g, \bar{g} \rangle$ is a pair of **P**-time function ensembles $g : \{0,1\}^n \times \{0,1\}^{\ell(n)} \to \{0,1\}^{\ell(n)}$ and $\bar{g} : \{0,1\}^n \times \{0,1\}^{\ell(n)} \to \{0,1\}^{\ell(n)}$, with the following additional properties. For each fixed $x \in \{0,1\}^n$, both $g_x(y)$ and $\bar{g}_x(y)$ as functions of $y \in \{0,1\}^{\ell(n)}$ are permutations. Furthermore, they are inverses of each other, i.e., for all $y \in \{0,1\}^{\ell(n)}$,

$$\bar{g}_x(g_x(y)) = g_x(\bar{g}_x(y)) = y.$$

The security parameter is $\|x\| = n$. The pair $\langle g, \bar{g} \rangle$ is a $\mathbf{S}(n)$-*secure pseudorandom invertible permutation generator* if $g_x(y)$ is a $\mathbf{S}(n)$-secure pseudorandom function generator. ♣

We use the following operator as the basic step in our construction of a pseudorandom invertible permutation generator.

Definition (operator $\mathcal{H}$): The *operator* $\mathcal{H}$ applied to a function $f_1 \in$ **Fnc**:$\{0,1\}^n \to \{0,1\}^n$ is a permutation

$$\mathcal{H}^{f_1} \in \mathbf{Perm}\text{:}\{0,1\}^{2n} \to \{0,1\}^{2n}$$

defined as follows. For all $z = \langle z_1, z_2 \rangle \in \{0,1\}^{2 \times n}$,

$$\mathcal{H}^{f_1}(z) = \langle z_2, z_1 \oplus f_1(z_2) \rangle.$$

For all $z = \langle z_1, z_2 \rangle \in \{0,1\}^{2 \times n}$, the inverse operator $\bar{\mathcal{H}}$ of $\mathcal{H}$ is defined as

$$\bar{\mathcal{H}}^{f_1}(z) = \langle z_2 \oplus f_1(z_1), z_1 \rangle.$$

More generally, let d be a positive integer, and let $f_1, \ldots, f_d$ all be in **Fnc**:$\{0,1\}^n \to \{0,1\}^n$. The operator $\mathcal{H}$ and the inverse operator $\bar{\mathcal{H}}$ with respect to $f_1, \ldots, f_d$ are defined inductively for all $d > 1$ as follows. For all $z \in \{0,1\}^{2 \times n}$,

$$\mathcal{H}^{f_1, \ldots, f_d}(z) = \mathcal{H}^{f_d}(\mathcal{H}^{f_1, \ldots, f_{d-1}}(z)),$$

and

$$\bar{\mathcal{H}}^{f_1, \ldots, f_d}(z) = \bar{\mathcal{H}}^{f_1}(\bar{\mathcal{H}}^{f_2, \ldots, f_d}(z)).$$

♣

In the above definition, if $f_1, \ldots, f_d$ are all **P**-time function ensembles then so are $\mathcal{H}^{f_1, \ldots, f_d}$ and $\bar{\mathcal{H}}^{f_1, \ldots, f_d}$.

Construction of an invertible permutation generator : Let $f : \{0,1\}^n \times \{0,1\}^n \to \{0,1\}^n$ be a pseudorandom function generator with security parameter n. Let $x \in \{0,1\}^{d \times n}$ and $y = \langle y_1, y_2 \rangle \in \{0,1\}^{2 \times n}$.

For any fixed integer $d \geq 1$, define the invertible permutation generator $\langle g^{(d)}, \bar{g}^{(d)} \rangle$ as follows:

$$g_x^{(d)}(y) = \mathcal{H}^{f_{x_1}, \ldots, f_{x_d}}(y)$$

and

$$\bar{g}_x^{(d)}(y) = \bar{\mathcal{H}}^{f_{x_1}, \ldots, f_{x_d}}(y).$$

The security parameter of $\langle g^{(d)}, \bar{g}^{(d)} \rangle$ is $\|x\| = dn$.

———∞———

The following adversary A shows that the invertible permutation generator $\langle g^{(1)}, \bar{g}^{(1)} \rangle$ is not at all pseudo-random. Fix $a = \langle a_1, a_2 \rangle \in \{0,1\}^{2 \times n}$ arbitrarily. A makes one oracle query with input a. Suppose $a' = \langle a_1', a_2' \rangle \in \{0,1\}^{2 \times n}$ is the answer received from the oracle query. The output of A is 1 if $a_1' = a_2$ and the output is 0 otherwise. Let $X \in_{\mathcal{U}} \{0,1\}^n$ and $F' \in_{\mathcal{U}} \mathbf{Fnc}{:}\{0,1\}^{2n} \to \{0,1\}^{2n}$. For any fixed $x \in \{0,1\}^n$, if $g_x^{(1)}(a) \in \{0,1\}^{2 \times n}$ is the value returned by the oracle query, then the output of A is 1 because $g_x^{(1)}(a)_1 = a_2$. Thus,

$$\Pr_X[A^{g_x^{(1)}} = 1] = 1.$$

On the other hand,

$$\Pr_{F'}[A^{F'} = 1] = 2^{-n},$$

because if F' is used to evaluate the oracle query then $F'(a)_1 \in_{\mathcal{U}} \{0,1\}^n$ independent of a. Thus, the success probability of A is $1 - 2^{-n}$.

In light of the DES construction, the natural conjecture is that the invertible permutation generator $\langle g^{(2)}, \bar{g}^{(2)} \rangle$ is pseudo-random. However, the following adversary A shows that this is not the case. Fix $\ell_1 \in \{0,1\}^n$, $\ell_2 \in \{0,1\}^n \setminus \{\ell_1\}$ and $r \in \{0,1\}^n$ arbitrarily. A makes two oracle queries. The first query is with input $a = \langle \ell_1, r \rangle$ and the second is with input $b = \langle \ell_2, r \rangle$. Let the answer returned from the first oracle query be $a' = \langle a_1', a_2' \rangle \in \{0,1\}^{2 \times n}$ and let the answer returned from the second oracle query be $b' = \langle b_1', b_2' \rangle \in \{0,1\}^{2 \times n}$. The output of A is 1 if $a_1' \oplus b_1' = \ell_1 \oplus \ell_2$ and the output is 0 otherwise. Let $X \in_{\mathcal{U}} \{0,1\}^{2 \times n}$ and $F' \in_{\mathcal{U}} \mathbf{Fnc}{:}\{0,1\}^{2n} \to \{0,1\}^{2n}$. For any fixed $x \in \{0,1\}^{2 \times n}$, if the oracle queries are evaluated using $g_x^{(2)}$ then the output of A is always 1. This is because $g_x^{(2)}(a)_1 = f_{x_1}(r) \oplus \ell_1$ and $g_x^{(2)}(b)_1 = f_{x_1}(r) \oplus \ell_2$ and thus

$$g_x^{(2)}(a)_1 \oplus g_x^{(2)}(b)_1 = \ell_1 \oplus \ell_2.$$

From this it follows that

$$\Pr_X[A^{g_x^{(2)}} = 1] = 1.$$

On the other hand,

$$\Pr_{F'}[A^{F'} = 1] = 2^{-n}.$$

This is because, for fixed a and b such that $a \neq b$, $F'(a)_1 \in_{\mathcal{U}} \{0,1\}^n$ and $F'(b)_1 \in_{\mathcal{U}} \{0,1\}^n$ are independent random variables, and thus

$$\Pr_{F'}[F'(a)_1 \oplus F'(b)_1 = \ell_1 \oplus \ell_2] = 2^{-n}.$$

Thus, the success probability of A is $1 - 2^{-n}$.

The Permutation Technical Theorem

Although neither $\langle g^{(1)}, \bar{g}^{(1)} \rangle$ nor $\langle g^{(2)}, \bar{g}^{(2)} \rangle$ is a pseudorandom invertible permutation generator, the Permutation Theorem on page 138 shows that $\langle g^{(3)}, \bar{g}^{(3)} \rangle$ is a pseudorandom invertible permutation generator. The primary technical component in the proof is the following theorem. The Permutation Technical Theorem is interesting in its own right: It says an easily computable and invertible permutation that looks random can be constructed from three easily computable random functions.

Permutation Technical Theorem : If A is an oracle adversary that produces a single bit and makes at most m oracle queries with inputs and outputs of length $2n$ then

$$\left| \Pr_{F_1, F_2, F_3}[A^{\mathcal{H}^{F_1, F_2, F_3}} = 1] - \Pr_{F_0}[A^{F_0} = 1] \right| \leq m^2/2^n,$$

where

$$F_1, F_2, F_3 \quad \in_{\mathcal{U}} \quad \mathbf{Fnc}{:}\{0,1\}^n \to \{0,1\}^n \text{ and}$$

$$F_0 \quad \in_{\mathcal{U}} \quad \mathbf{Fnc}{:}\{0,1\}^{2n} \to \{0,1\}^{2n}.$$

PROOF: Let p_0 and p_1 be defined as follows:

$$p_0 \quad = \quad \Pr_{F_0}[A^{F_0} = 1],$$

$$p_1 \quad = \quad \Pr_{F_1, F_2, F_3}[A^{\mathcal{H}^{F_1, F_2, F_3}} = 1].$$

We prove that $|p_0 - p_1| \leq m^2/2^n$. Let

$$X, Y, Z \in_{\mathcal{U}} \{0,1\}^{m \times n},$$

where X_i, Y_i and Z_i are to be used to produce the answer to the ith oracle call. We consider a probability distribution defined by

$$\langle X, Y, Z \rangle.$$

Let $x, y, z \in \{0, 1\}^{m \times n}$. We describe later two algorithms, B and C, for computing the answers to the oracle queries of A. The algorithm B is of the form

$$B(x, y, z) = \langle B_1(x, y, z), \ldots, B_m(x, y, z) \rangle,$$

where $B_i(x, y, z)$ is an algorithm for computing the ith oracle call. The algorithm C is of the form

$$C(x, y, z) = \langle C_1(x, y, z), \ldots, C_m(x, y, z) \rangle,$$

where again $C_i(x, y, z)$ is an algorithm for computing the ith oracle call. We prove that

$$\Pr_{X,Y,Z}[A^{B(X,Y,Z)} = 1] = p_1,$$

$$\Pr_{X,Y,Z}[A^{C(X,Y,Z)} = 1] = p_0$$

and

$$\Pr_{X,Y,Z}[A^{B(X,Y,Z)} \neq A^{C(X,Y,Z)}] \leq m^2/2^n.$$

This implies that $|p_0 - p_1| \leq m^2/2^n$, as desired.

We assume that A never repeats an input to an oracle call. This assumption is without loss of generality because we can simulate any adversary which repeats inputs by another adversary which remembers answers to all past queries and hence does not have to repeat inputs. The simulating adversary has exactly the same number of oracle queries. When the original adversary makes an oracle query the simulating adversary first checks to see if the current input to the oracle query has occurred previously. If the current input is different from all previous inputs to oracle queries, the simulating adversary uses the current input to make the oracle query and returns the answer to the original adversary. If the current input is the same as a previous input to an oracle query, the simulating adversary computes a new input that is different from all previous inputs (including the current input) and makes the oracle query with the new input. The simulating adversary then looks up the answer corresponding to the query with the current input (this query was made previously) and returns this as the answer to the original adversary for this query (even though the simulating adversary made this query with the new input).

Before defining $B(x, y, z)$ and $C(x, y, z)$, to simplify the proof we define an intermediate way of computing the oracle queries,

$$B'(x, y, z) = \langle B_1(x, y, z), \ldots, B_m(x, y, z) \rangle.$$

In the following description, the input to the ith oracle query is

$$\langle L_i(x, y, z), R_i(x, y, z) \rangle,$$

where $L_i(x, y, z) \in \{0, 1\}^n$ and $R_i(x, y, z) \in \{0, 1\}^n$.

Oracle computation $B_i'(x, y, z)$ on input $\langle L_i(x, y, z), R_i(x, y, z) \rangle$:

Let $u_i(x, y, z) = \min\{j \in \{1, \ldots, i\} : R_i(x, y, z) = R_j(x, y, z)\}$.

$\quad \alpha_i(x, y, z) = L_i(x, y, z) \oplus x_{u_i(x, y, z)}$.

Let $v_i(x, y, z) = \min\{j \in \{1, \ldots, i\} : \alpha_i(x, y, z) = \alpha_j(x, y, z)\}$.

$\quad \beta_i(x, y, z) = R_i(x, y, z) \oplus y_{v_i(x, y, z)}$.

Let $w_i(x, y, z) = \min\{j \in \{1, \ldots, i\} : \beta_i(x, y, z) = \beta_j(x, y, z)\}$.

$\quad \gamma_i(x, y, z) = \alpha_i(x, y, z) \oplus z_{w_i(x, y, z)}$.

Output $\langle \beta_i(x, y, z), \gamma_i(x, y, z) \rangle$

Formally $B'(x, y, z)$ is also a function of the adversary A, but we suppress this dependence in the notation. It is not hard to verify that

$$\Pr_{X, Y, Z}[A^{B'(X, Y, Z)} = 1] = p_1.$$

We now describe $B(x, y, z)$. The input to the ith oracle query is

$$\langle L_i(x, y, z), R_i(x, y, z) \rangle,$$

where $L_i(x, y, z) \in \{0, 1\}^n$ and $R_i(x, y, z) \in \{0, 1\}^n$.

Oracle computation $B_i(x, y, z)$ on input $\langle L_i(x, y, z), R_i(x, y, z) \rangle$:

Let $u_i(x, y, z) = \min\{j \in \{1, \ldots, i\} : R_i(x, y, z) = R_j(x, y, z)\}$.

$\quad \alpha_i(x, y, z) = L_i(x, y, z) \oplus x_{u_i(x, y, z)}$.

Let $v_i(x, y, z) = \min\{j \in \{1, \ldots, i\} : \alpha_i(x, y, z) = \alpha_j(x, y, z)\}$.

$\pi_i(x, y, z) = y_i \oplus R_i(x, y, z)$.

$\beta_i(x, y, z) = R_i(x, y, z) \oplus \pi_{v_i(x,y,z)}(x, y, z)$.

Let $w_i(x, y, z) = \min\{j \in \{1, \ldots, i\} : \beta_i(x, y, z) = \beta_j(x, y, z)\}$.

$\rho_i(x, y, z) = z_i \oplus \alpha_i(x, y, z)$.

$\gamma_i(x, y, z) = \alpha_i(x, y, z) \oplus \rho_{w_i(x,y,z)}(x, y, z)$.

Output $\langle \beta_i(x, y, z), \gamma_i(x, y, z) \rangle$

Claim : $\Pr_{X,Y,Z}[A^{B'(X,Y,Z)} = 1] = \Pr_{X,Y,Z}[A^{B(X,Y,Z)} = 1] = p_1$.

PROOF: Let

$$\pi(x, y, z) = \langle \pi_1(x, y, z), \ldots, \pi_m(x, y, z) \rangle$$

and

$$\rho(x, y, z) = \langle \rho_1(x, y, z), \ldots, \rho_m(x, y, z) \rangle$$

be defined with respect to B. By induction on i:

- The value of $R_i(x, y, z)$ only depends on x_j, y_j, and z_j for $j < i$.

- The value of $\alpha_i(x, y, z)$ only depends on x_j, y_j, and z_j for $j < i$ and on x_i.

Thus, X_i,

$$\pi_i(X, Y, Z) = Y_i \oplus R_i(X, Y, Z)$$

and

$$\rho_i(X, Y, Z) = Z_i \oplus \alpha_i(X, Y, Z)$$

are independent of each other and independent of

$$X_1, \ldots, X_{i-1}, Y_1, \ldots, Y_{i-1}, Z_1, \ldots, Z_{i-1}.$$

This shows that X, $\pi(X, Y, Z)$ and $\rho(X, Y, Z)$ are uniformly and independently distributed in $\{0, 1\}^{m \times n}$. The claim follows because the role of $\langle X, \pi(X, Y, Z), \rho(X, Y, Z) \rangle$ with respect to B is the same as the role of $\langle X, Y, Z \rangle$ with respect to B'. ∎

We rewrite $B(x, y, z)$ in a more compact fashion (eliminating references to $\pi_1, \ldots, \pi_m$ and $\rho_1, \ldots, \rho_m$) that is convenient for the remainder of the proof.

Oracle computation $B_i(x, y, z)$ on input $\langle L_i(x, y, z), R_i(x, y, z) \rangle$:

Let $u_i(x, y, z) = \min\{j \in \{1, \ldots, i\} : R_i(x, y, z) = R_j(x, y, z)\}$.

$\qquad \alpha_i(x, y, z) = L_i(x, y, z) \oplus x_{u_i(x,y,z)}$.

Let $v_i(x, y, z) = \min\{j \in \{1, \ldots, i\} : \alpha_i(x, y, z) = \alpha_j(x, y, z)\}$.

$\qquad \beta_i(x, y, z) = R_i(x, y, z) \oplus R_{v_i(x,y,z)}(x, y, z) \oplus y_{v_i(x,y,z)}$.

Let $w_i(x, y, z) = \min\{j \in \{1, \ldots, i\} : \beta_i(x, y, z) = \beta_j(x, y, z)\}$.

$\qquad \gamma_i(x, y, z) = \alpha_i(x, y, z) \oplus \alpha_{w_i(x,y,z)}(x, y, z) \oplus z_{w_i(x,y,z)}$.

Output $\langle \beta_i(x, y, z), \gamma_i(x, y, z) \rangle$.

$C(x, y, z)$ is defined the same way as $B(x, y, z)$, except that the output of $C_i(x, y, z)$ is $\langle y_i, z_i \rangle$. Because we assumed that A never repeats the same input to an oracle computation,

$$\Pr_{X,Y,Z}[A^{C(X,Y,Z)} = 1] = p_0.$$

The outline of the proof is to define triple $\langle x, y, z \rangle$ to be *bad* if $A^{B(x,y,z)} \neq A^{C(x,y,z)}$, and show the probability that $\langle X, Y, Z \rangle$ is bad is at most $m^2/2^n$.

Definition (bad with respect to y): $\langle x, y, z \rangle$ is *bad* with respect to y, if there are i, j, $1 \leq i < j \leq m$ such that $y_i = y_j$. ♣

The probability $\langle X, Y, Z \rangle$ is bad with respect to Y is at most $m^2/2^{n+1}$.

Definition (bad with respect to x): $\langle x, y, z \rangle$ is *bad* with respect to x if, with respect to the computation of $A^{C(x,y,z)}$, there are i, j, $1 \leq i < j \leq m$ such that

$$L_j(x, y, z) \oplus x_{u_j(x,y,z)} = L_i(x, y, z) \oplus x_{u_i(x,y,z)}.$$

♣

Claim : The probability $\langle X, Y, Z \rangle$ is bad with respect to X is at most $m^2/2^{n+1}$.

PROOF: By the way $A^{C(x,y,z)}$ is computed, for all $i = 1, \ldots, m$, y and z completely determine the output of $C_i(x, y, z)$, and thus y and

z also determine everything about the computation except for the internal computations of C_i. In particular, y and z also completely determine the inputs to the oracle queries $L_1(x, y, z), \ldots, L_m(x, y, z)$ and $R_1(x, y, z), \ldots, R_m(x, y, z)$, and these inputs in turn determine the values of $u_1(x, y, z), \ldots, u_m(x, y, z)$. This implies that for fixed j, y and z, the values of $L_j(X, y, z)$, $R_j(X, y, z)$ and $u_j(X, y, z)$ are fixed independently of X. We show below that, for $i \neq j$ and for fixed y and z, the probability with respect to X that the event

$$L_j(X, y, z) \oplus X_{u_j(X,y,z)} = L_i(X, y, z) \oplus X_{u_i(X,y,z)}$$

occurs is at most 2^{-n}.

- If $u_i(X, y, z) = u_j(X, y, z)$ then

$$L_j(X, y, z) \oplus X_{u_j(X,y,z)} = L_i(X, y, z) \oplus X_{u_i(X,y,z)}$$

is the same as

$$L_j(X, y, z) = L_i(X, y, z).$$

But this event never occurs, because $u_i(X, y, z) = u_j(X, y, z)$ also implies that $R_i(X, y, z) = R_j(X, y, z)$, and because A does not repeat inputs to oracle queries, this implies that

$$L_i(X, y, z) \neq L_j(X, y, z).$$

- If $u_i(X, y, z) \neq u_j(X, y, z)$ then, letting $i' = u_i(X, y, z)$, $j' = u_j(X, y, z)$ and $a = L_i(X, y, z) \oplus L_j(X, y, z)$,

$$L_j(X, y, z) \oplus X_{u_j(X,y,z)} = L_i(X, y, z) \oplus X_{u_i(X,y,z)}$$

is the same as

$$X_{i'} \oplus X_{j'} = a.$$

But this event occurs with probability 2^{-n}.

From this it follows that the probability that any of the $\binom{m}{2}$ events occurs is at most $m^2/2^{n+1}$. This complete the proof of the claim. ∎

To summarize so far, we have shown that the probability $\langle X, Y, Z \rangle$ is bad with respect to either X or Y is at most $m^2/2^n$. Below, we show that $A^{B(x,y,z)} = A^{C(x,y,z)}$ whenever $\langle x, y, z \rangle$ is not bad with respect to both x and y. Putting these two facts together shows that

$$\Pr_{X,Y,Z}[A^{B(X,Y,Z)} \neq A^{C(X,Y,Z)}] \leq m^2/2^n,$$

which will complete the proof of the theorem.

Definition (preserving): $\langle x, y, z \rangle$ is *preserving* if, with respect to the computation of $A^{C(x,y,z)}$, for all $i = 1, \ldots, m$, $v_i(x, y, z) = i$ and $w_i(x, y, z) = i$. ♣

Claim : If $\langle x, y, z \rangle$ is not bad with respect to both x and y then $\langle x, y, z \rangle$ is preserving.

PROOF: If $\langle x, y, z \rangle$ is not bad with respect to x then it can be easily verified that for all $i = 1, \ldots, m$, $v_i(x, y, z) = i$ in the computation of $C_i(x, y, z)$. Therefore, for all $i = 1, \ldots, m$, $\beta_i(x, y, z) = y_i$. If $\langle x, y, z \rangle$ is not bad with respect to y then, because for all $i = 1, \ldots, m$, $\beta_i(x, y, z) = y_i$, it follows that for all $i = 1, \ldots, m$, $\gamma_i(x, y, z) = z_i$. This complete the proof of the claim. ■

Claim : If $\langle x, y, z \rangle$ is preserving then $A^{B(x,y,z)} = A^{C(x,y,z)}$.

PROOF: The output of $C_i(x, y, z)$ is $\langle y_i, z_i \rangle$. The output of $B_i(x, y, z)$ is $\langle \beta_i(x, y, z), \gamma_i(x, y, z) \rangle$. We prove, by induction on i, that if $\langle x, y, z \rangle$ is preserving then the computation of A using $B_i(x, y, z)$ to compute the oracle queries is exactly the same as the computation of A using $C_i(x, y, z)$ to compute the oracle queries. Suppose the computation is exactly the same up to the ith oracle call. Since the internal computations of $B_i(x, y, z)$ and $C_i(x, y, z)$ are exactly the same, and since $\langle x, y, z \rangle$ is preserving implies that $\beta_i(x, y, z) = y_i$ and $\gamma_i(x, y, z) = z_i$, it follows that the output of $B_i(x, y, z)$ and $C_i(x, y, z)$ are exactly the same. This complete the proof of the claim. ■

From the preceding two claims it follows that if $\langle x, y, z \rangle$ is not bad with respect to both x and y then $A^{B(x,y,z)} = A^{C(x,y,z)}$. This completes the proof of the Permutation Technical Theorem. ■

Even though $\mathcal{H}^{F_1, F_2, F_3}$ is a permutation, the Permutation Technical Theorem (page 131) shows that it is indistinguishable from a random function F_0. Defining indistinguishability with respect to a random function as opposed to a random permutation is not important, as the following exercise shows.

Exercise 51 : Let

$$F_0 \in_{\mathcal{U}} \mathbf{Perm}{:}\{0,1\}^n \rightarrow \{0,1\}^n$$

and

$$F_1 \in_{\mathcal{U}} \mathbf{Fnc}{:}\{0,1\}^n \rightarrow \{0,1\}^n.$$

Show that any adversary A that makes at most m queries to an oracle has success probability at most $m^2/2^n$ for distinguishing F_0 and F_1. ♠

Lecture 14

Overview

We show how to construct a pseudorandom invertible permutation generator. We define and construct a super pseudorandom invertible permutation generator. We use these constructions to design secure block private key cryptosystems.

The Permutation Theorem

We show how to construct a pseudorandom invertible permutation generator from a pseudorandom function generator. Let

$$f : \{0,1\}^n \times \{0,1\}^n \to \{0,1\}^n$$

be a pseudorandom function generator and let $\langle g^{(3)}, \bar{g}^{(3)} \rangle$ be the invertible permutation generator constructed from f as described on page 129.

Permutation Theorem : If f is a pseudorandom function generator then $\langle g^{(3)}, \bar{g}^{(3)} \rangle$ is a pseudorandom invertible permutation generator. The reduction is linear-preserving.

PROOF: Let

$$F_0 \quad \in_{\mathcal{U}} \quad \mathbf{Fnc}{:}\{0,1\}^{2n} \to \{0,1\}^{2n},$$

$$F_1, F_2, F_3 \quad \in_{\mathcal{U}} \quad \mathbf{Fnc}{:}\{0,1\}^n \to \{0,1\}^n \text{ and}$$

$$X \quad \in_{\mathcal{U}} \quad \{0,1\}^{3 \times n}.$$

Suppose A is an adversary for $\langle g^{(3)}, \bar{g}^{(3)} \rangle$ with success probability $\delta(n)$ and run time $T(n)$. Let

$$
\begin{aligned}
p_0 &= \Pr[A^{F_0} = 1], \\
p_1 &= \Pr[A^{\mathcal{H}^{F_1, F_2, F_3}} = 1], \\
p_2 &= \Pr[A^{\mathcal{H}^{f_{X_1}, F_2, F_3}} = 1], \\
p_3 &= \Pr[A^{\mathcal{H}^{f_{X_1}, f_{X_2}, F_3}} = 1], \\
p_4 &= \Pr[A^{g_X^{(3)}} = 1] = \Pr[A^{\mathcal{H}^{f_{X_1}, f_{X_2}, f_{X_3}}} = 1].
\end{aligned}
$$

By definition of the success probability of A, $|p_4 - p_0| = \delta(n)$. A can make at most $T(n)$ oracle queries in time $T(n)$, and thus the Permutation Technical Theorem shows that $|p_1 - p_0| \leq T(n)^2/2^n$ and we assume this is at most $\delta(n)/2$ without much loss in generality. From this it follows that

$$|p_4 - p_1| \geq \delta(n)/2.$$

The oracle adversary S^A makes queries to some function

$$f' \in \mathbf{Fnc}{:}\{0,1\}^n \to \{0,1\}^n.$$

Thus, S can be thought of as an oracle adversary that makes two kinds of oracle queries, queries to f' and queries to A, and we denote this by $S^{f',A}$. The situation is even a bit more complicated than that: S simulates the computation of A, where A is also an oracle adversary. Whenever A makes an oracle call, S takes the input to the oracle call that A makes and produces an output for the oracle call, passes this back to A and the simulation continues. We describe exactly how S works below.

Let $V \in_\mathcal{U} \{0,1\}^n$ and $F' \in_\mathcal{U} \mathbf{Fnc}{:}\{0,1\}^n \to \{0,1\}^n$. S simulates A to distinguish between when f' is chosen according to f_V and when f' is chosen according to F'. At a high level, $S^{f',A}$ randomly chooses $i \in_\mathcal{U} \{1,2,3\}$, $x \in_\mathcal{U} \{0,1\}^{2 \times n}$ and $f_2, f_3 \in_\mathcal{U} \mathbf{Fnc}{:}\{0,1\}^n \to \{0,1\}^n$.

- If $i = 1$ then $S^{f',A}$ simulates A, where the oracle queries A makes are computed using $\mathcal{H}^{f',f_2,f_3}$.

- If $i = 2$ then $S^{f',A}$ simulates A, where the oracle queries A makes are computed using $\mathcal{H}^{f_{x_1},f',f_3}$.

- If $i = 3$ then $S^{f',A}$ simulates A, where the oracle queries A makes are computed using $\mathcal{H}^{f_{x_1},f_{x_2},f'}$.

The final output of $S^{f',A}$ is the output of A after the simulation.

Let m be an upper bound on the number of oracle queries A makes. $S^{f',A}$ doesn't really randomly choose $f_2, f_3 \in_\mathcal{U} \mathbf{Fnc}{:}\{0,1\}^n \to \{0,1\}^n$. Instead, $S^{f',A}$ random chooses $a, b \in_\mathcal{U} \{0,1\}^{m \times n}$, and uses a to simulate the at most m queries to f_2 and b to simulate the at most m queries to f_3. On the ith oracle query to f_2, if the input is the same as some previous input to f_2 then $S^{f',A}$ gives the same answer as before, and otherwise the answer is a_i. $S^{f',A}$ uses b in a similar way to simulate the queries to f_3.

The success probability of S^A is equal to

$$\left| \Pr_V[S^{f_V,A} = 1] - \Pr_{F'}[S^{F',A} = 1] \right| = |p_4 - p_1|/3 \geq \delta(n)/6.$$

The total time it takes for $S^{f',A}$ to simulate A is not much more than the run time of A. ∎

Exercise 52 : Let $f : \{0,1\}^n \times \{0,1\}^n \to \{0,1\}^n$ be a pseudorandom function generator. Define invertible permutation generator $\langle g, \bar{g} \rangle$ as

$$g_x(z) = \mathcal{H}^{f_x, f_x, f_x}(z),$$

and

$$\bar{g}_x(z) = \bar{\mathcal{H}}^{f_x, f_x, f_x}(z),$$

where $z \in \{0,1\}^{2 \times n}$ and operator $\mathcal{H}$ is defined on page 129. Prove or disprove that $\langle g, \bar{g} \rangle$ is a pseudorandom invertible permutation generator. ♠

A Block Cryptosystem Without Explicit Indexing

In the block private key cryptosystem defined on page 118 a unique index was associated with each message. In many practical applications, especially when the length of each message is fairly long, it is highly unlikely that the exact same message will ever be sent twice using the same private key. It is simpler to let the encryption depend only on the message itself, as is the case for DES.

Definition (block private key cryptosystem without explicit indexing): A *block private key cryptosystem without explicit indexing* consists of the following.

(initialization): All parties exchange information on private lines to establish a private key $x \in \{0,1\}^n$. All parties store x in their respective private memories, and x is considered the security parameter of the protocol.

(message sending): Let

$$E : \{0,1\}^n \times \{0,1\}^{q(n)} \to \{0,1\}^{k(n)}$$

and

$$D : \{0,1\}^n \times \{0,1\}^{k(n)} \to \{0,1\}^{q(n)}$$

be **P**-time function ensembles. E and D have the property that, for all $x \in \{0,1\}^n$ and for all $m \in \{0,1\}^{q(n)}$,

$$D_x(E_x(m)) = m.$$

A party sends a message $m \in \{0,1\}^{q(n)}$ by privately computing $e = E_x(m)$ and sending this on a public line. Upon receiving e, another party can recover m by computing $D_x(e)$ using the private memory device, storing the result presumably in private memory. All messages sent using the same private key must be distinct. ♣

Definition (chosen plaintext attack for a block system without explicit indexing):

Let $M : \{0,1\}^{\log(p(n))} \times \{0,1\}^{s(n)} \times \{0,1\}^{p(n)k(n)} \rightarrow \{0,1\}^{q(n)}$,
$\quad P : \{0,1\}^{s(n)} \times \{0,1\}^{p(n)k(n)} \rightarrow \{0,1\}^{2q(n)}$,
$\quad A : \{0,1\}^{s(n)} \times \{0,1\}^{p(n)k(n)} \times \{0,1\}^{k(n)} \rightarrow \{0,1\}$ be adversaries.

The attack works as follows.

(choose a private key): Choose a private key $x \in_{\mathcal{U}} \{0,1\}^n$.

(chosen plaintext attack): Choose $r \in_{\mathcal{U}} \{0,1\}^{s(n)}$. For $j = 1, \ldots, p(n)$, phase j works as follows. Let

$$e = \langle E_x(m_1), \ldots, E_x(m_{j-1}) \rangle$$

be the concatenation of the encryptions of the first $j - 1$ message blocks padded out with zeroes to a string of length $p(n)k(n)$. Then, $m_j = M(j, r, e)$. At the end of the $p(n)$ phases, let $m = \langle m_1, \ldots, m_{p(n)} \rangle$ be all the message blocks produced by M, and let $e = \langle E_x(m_1), \ldots, E_x(m_{p(n)}) \rangle$ be the encryption of m.

(choose a private message block): Let $\langle m^0, m^1 \rangle = P(r, e)$ be the pair of message blocks produced by P. It is required that neither m^0 nor m^1 is among the message blocks in m. Choose $b \in_{\mathcal{U}} \{0,1\}$ privately, let $m' = m^b$ be the privately chosen message, and let $e' = E_x(m')$ be the encryption of m'.

(predict the bit): The success probability of the adversary is

$$\delta(n) = |\mathrm{E}[\overline{A(r, e, e')} \cdot \overline{b}]|.$$

The run time $T(n)$ of the overall adversary includes the time to compute M, P and A. The block cryptosystem is $\mathbf{S}(n)$-*secure against chosen plaintext attack* if every adversary has time-success ratio at least $\mathbf{S}(n)$. ♣

Note that when P generates the two possible message blocks m^0 and m^1, P implicitly knows all message blocks m generated during the chosen plaintext attack, since these can be computed based on r and e using

M. Similarly, when A is trying to predict b, A implicitly knows m, m^0 and m^1, since these can be computed based on r and e using M and P.

The following exercise shows that if the condition that P generate two message blocks that are distinct from all previous message blocks is removed then the encryption system can be insecure even when using a perfect encryption function.

Exercise 53 : Suppose that P is allowed to produce a pair of message blocks that are equal to a message block produced by M in the definition of chosen plaintext attack given above. Describe a P, M and A that have a constant success probability, independent of the security of the encryption function used. ♠

Because the notion of a pseudorandom invertible permutation generator was inspired by the specific example of DES, it is not surprising that a secure block cryptosystem of this type can be easily constructed from a pseudorandom invertible permutation generator.

Construction of a block cryptosystem from a pseudorandom invertible permutation generator : Let $\langle g, \bar{g} \rangle$ be a pseudorandom invertible permutation generator with $g : \{0,1\}^n \times \{0,1\}^n \to \{0,1\}^n$ and $\bar{g} : \{0,1\}^n \times \{0,1\}^n \to \{0,1\}^n$. Define a block private key cryptosystem as follows. Let $x \in \{0,1\}^n$ be the private key. The encryption of message $m \in \{0,1\}^n$ is $e = g_x(m)$. The decryption of e is $\bar{g}_x(e)$.

————∞————

The following exercise is analogous to Exercise 48 (page 126).

Exercise 54 : Prove that if $\langle g, \bar{g} \rangle$ is a pseudorandom invertible permutation generator then the block private key cryptosystem just described is secure against chosen plaintext attack, where the reduction is linear-preserving. ♠

super pseudorandom invertible permutation generator

One other kind of attack often considered besides chosen plaintext attack is chosen ciphertext attack. This is where the adversary is given access to the decryption device (once again treated as an oracle) which she/he may use to decrypt interactively chosen encryptions at will. The intuition for a super pseudorandom invertible permutation generator is that it is secure against simultaneous chosen plaintext and chosen ciphertext attack.

Definition (super pseudorandom invertible permutation generator): Let $\langle g, \bar{g} \rangle$ be an invertible permutation generator with g : $\{0,1\}^n \times \{0,1\}^n \to \{0,1\}^n$ and $\bar{g} : \{0,1\}^n \times \{0,1\}^n \to \{0,1\}^n$. Let adversary A be an oracle adversary that makes two kinds of oracle queries, forward and inverse. Let $f' \in \mathbf{Perm}:\{0,1\}^n \to \{0,1\}^n$ be a permutation and $\bar{f}'$ the inverse permutation of f'. The oracle queries made by $A^{f', \bar{f}'}$ are computed using f' for the forward queries and using $\bar{f}'$ for the inverse queries. The output of $A^{f', \bar{f}'}$ is a single bit. Let $X \in_{\mathcal{U}} \{0,1\}^n$, $F' \in_{\mathcal{U}} \mathbf{Perm}:\{0,1\}^n \to \{0,1\}^n$ and let $\bar{F}'$ be the inverse permutation of F'. The success probability of A for $\langle g, \bar{g} \rangle$ is

$$\delta(n) = |\Pr_X[A^{gx, \bar{g}x} = 1] - \Pr_{F'}[A^{F', \bar{F}'} = 1]|.$$

Then, $\langle g, \bar{g} \rangle$ is a $\mathbf{S}(n)$-*secure super pseudorandom invertible permutation generator* if every adversary has time-success ratio at least $\mathbf{S}(n)$. ♣

Exercise 55 : Let $\langle g^{(3)}, \bar{g}^{(3)} \rangle$ be the invertible permutation generator constructed from a pseudorandom function generator f as described on page 129. The Permutation Theorem (page 138) shows that $\langle g^{(3)}, \bar{g}^{(3)} \rangle$ is a pseudorandom invertible permutation generator. Show that $\langle g^{(3)}, \bar{g}^{(3)} \rangle$ is definitely not a super pseudorandom invertible permutation generator. ♠

Exercise 56 : Let $\langle g^{(4)}, \bar{g}^{(4)} \rangle$ be the invertible permutation generator constructed from a pseudorandom function generator f as described on page 129. Prove that $\langle g^{(4)}, \bar{g}^{(4)} \rangle$ is a super pseudorandom invertible permutation generator, where the reduction is linear-preserving. ♠

Simultaneous Plaintext and Ciphertext Attack

The intuition behind the attack is that there is a party who is willing to encrypt message blocks generated by an adversary and reveal the corresponding encryptions. Furthermore, the party is willing to decrypt encryptions generated by the adversary and reveal the corresponding message to the adversary. At some point in time the party generates an important message privately that is not revealed to the adversary, encrypts this important message, and sends it over a public line where it is intercepted by the adversary. Intuitively, the attack is secure if the adversary cannot even predict one bit of information about the private message.

The attack allowed by an adversary is quite strong, and thus security against this type of attack is correspondingly strong.

Definition (simultaneous attack): Suppose the message blocks and their encryptions are both n bits each.

Let $M : \{0,1\}^{\log(p(n))} \times \{0,1\}^{s(n)} \times \{0,1\}^{p(n)n} \to \{0,1\}^{n+1}$,
   $P : \{0,1\}^{s(n)} \times \{0,1\}^{p(n)n} \to \{0,1\}^{2n}$,
   $A : \{0,1\}^{s(n)} \times \{0,1\}^{p(n)n} \times \{0,1\}^{n} \to \{0,1\}$ be adversaries.

The attack works as follows.

(choose a private key): Choose a private key $x \in_{\mathcal{U}} \{0,1\}^{n}$.

(simultaneous attack): Choose $r \in_{\mathcal{U}} \{0,1\}^{s(n)}$. For $j = 1, \ldots, p(n)$, phase j works as follows. Let $\beta = \langle \beta_1, \ldots, \beta_{j-1} \rangle$ consist of $j-1$ strings of length n each defined in previous phases, padded out with zeroes to a string of length $p(n)n$. Then, $\langle a_j, \alpha_j \rangle = M(j, r, \beta)$, where $a_j \in \{0,1\}$ indicates whether to try and encrypt or decrypt $\alpha_j \in \{0,1\}^{n}$. Then, $\beta_j = E_x(\alpha_j)$ if $a_j = 0$ and $\beta_j = D_x(\alpha_j)$ if $a_j = 1$. At the end of the $p(n)$ phases, let $\beta = \langle \beta_1, \ldots, \beta_{p(n)} \rangle$, and let $m = \langle m_1, \ldots, m_{p(n)} \rangle$ be the message blocks and $e = \langle e_1, \ldots, e_{p(n)} \rangle$ be the corresponding encryptions generated either directly or indirectly by M, i.e., $m_j = \alpha_j$ and $e_j = \beta_j$ if $a_j = 0$ and $m_j = \beta_j$ and $e_j = \alpha_j$ if $a_j = 1$.

(choose a private message block): Let $\langle m^0, m^1 \rangle = P(r, \beta)$ be a pair of message blocks generated by P. It is required that neither m^0 nor m^1 is among the message blocks in m. Choose $b \in_{\mathcal{U}} \{0,1\}$ privately, let $m' = m^b$ be the privately chosen message, and let $e' = E_x(m')$ be the encryption of m'.

(predict the bit): The success probability of the adversary is

$$\delta(n) = |\mathrm{E}[\overline{A(r, \beta, e') \cdot b}]|.$$

The run time $T(n)$ of the overall adversary includes the time to compute M, P and A. The block cryptosystem is $\mathbf{S}(n)$-*secure against simultaneous attack* if every adversary has time-success ratio at least $\mathbf{S}(n)$.   ♣

Note that when P generates the two possible message blocks m^0 and m^1, P implicitly knows all message blocks m and corresponding encryption blocks e generated during the simultaneous attack, since these can be computed based on r and β using M. Similarly, when A is trying to predict b, A implicitly knows m, e, m^0 and m^1, since these can be computed based on r and β using M and P.

An imaginable way an adversary could partially enact this type of attack is the following. The party goes off to lunch, leaving its encryption and decryption devices unprotected for use by the adversary for a period of time. When the party returns from lunch, the party sends the important message in encrypted form. At the end of the attack, the adversary is trying to predict a bit of information only about the important message.

The construction of a block cryptosystem secure against simultaneous attack consists of using the construction given on page 142 of a block private key cryptosystem based on a pseudorandom invertible permutation generator, only using a super pseudorandom invertible permutation generator instead of a pseudorandom invertible permutation generator.

The following exercise is analogous to Exercise 48 (page 126) and to Exercise 54 (page 142).

Exercise 57 : Consider a block cryptosystem constructed from a super pseudorandom invertible permutation generator $\langle g, \bar{g} \rangle$ as described above. Prove that if $\langle g, \bar{g} \rangle$ is a super pseudorandom invertible permutation generator then the block cryptosystem is secure against simultaneous attack, where the reduction is linear-preserving. ♠

Lecture 15

Overview

We introduce trapdoor one-way functions, one-way predicates and trapdoor one-way predicate, and based on this design cryptosystems without an initial communication using a private line.

Trapdoor Functions

We now introduce a stronger form of a one-way function that has additional useful properties.

Definition (trapdoor one-way function): Let $\mathcal{D}_n : \{0,1\}^{r(n)} \to \{0,1\}^{m(n)+\ell(n)}$ be a **P**-samplable probability ensemble. We call $\mathcal{D}_n$ the *key generation distribution*. Let $\langle x, z \rangle \in \{0,1\}^{m(n)} \times \{0,1\}^{\ell(n)}$ be a possible output of $\mathcal{D}_n$. We call x the *trapdoor key* and z the *public key*. Let $f : \{0,1\}^{\ell(n)} \times \{0,1\}^n \to \{0,1\}^{k(n)}$ be a **P**-time function ensemble, where the first input is the public key and the second input is private. For fixed z, $f_z(y)$ as a function y maps $\{0,1\}^n$ to $\{0,1\}^{k(n)}$. The following properties hold:

(invertible with the trapdoor key) Let $y \in \{0,1\}^n$. There is a **P**-time function ensemble that on input $\langle z, f_z(y), x \rangle$ produces $y' \in \{0,1\}^n$ such that $f_z(y') = f_z(y)$.

(one-way function) Let $\langle X, Z \rangle \in_{\mathcal{D}_n} \{0,1\}^{m(n)} \times \{0,1\}^{\ell(n)}$ and $Y \in_{\mathcal{U}} \{0,1\}^n$. We view both X and Y as private, and thus the security parameter $\mathbf{s}(n)$ is $m(n)+n$. Let $A : \{0,1\}^{\ell(n)} \times \{0,1\}^{k(n)} \to \{0,1\}^n$ be an adversary. Define the success probability of A as

$$\delta(n) = \Pr[f_Z(A(Z, f_Z(Y))) = f_Z(Y)].$$

Then, f is a $\mathbf{S}(\mathbf{s}(n))$-*secure trapdoor one-way function* if every adversary has time-success ratio at least $\mathbf{S}(\mathbf{s}(n))$. ♣

This definition is similar to the definition of a one-way function, except that instead of defining a single function from $\{0,1\}^n$ to $\{0,1\}^{k(n)}$ for each value of n, it is a family of functions indexed by a public key $z \in \{0,1\}^{\ell(n)}$. Although each function in the family is hard on average to invert given only the public key z, it is easy to invert given the trapdoor key x that is produced along with z by the key generation distribution. A trapdoor one-way function is easily seen to be one-way function, but it is not known if a trapdoor one-way function can be constructed from any one-way function.

Definition (trapdoor one-way permutation): A $\mathbf{S}(\mathbf{s}(n))$-*secure trapdoor one-way permutation* is a $\mathbf{S}(\mathbf{s}(n))$-secure trapdoor one-way function $f_z(y)$ with the additional property that, for each fixed $z \in \{0,1\}^{\ell(n)}$, f_z is a permutation.   ♣

Root extraction problem : The root extraction problem (page 17) is an example of a conjectured trapdoor one-way function. Let $\langle p, q \rangle$ be a pair of primes of length n each, and let e be a positive integer. Let $x = \langle p, q \rangle$ be the trapdoor key and $z = \langle pq, e \rangle$ be the public key. Define $f_z(y) = y^e \bmod pq$, where $y \in \mathcal{Z}_z^*$. Recall that an inverse of $f_z(y)$ can be computed in $n^{\mathcal{O}(1)}$ time given e and the factorization $\langle p, q \rangle$ of pq. The distribution on which this function is conjectured to be hard to invert is when the pair of primes $\langle P, Q \rangle$ is randomly chosen so that $Z = PQ$ is hard to factor on average, $e \geq 2$ is fixed, and $Y \in_{\mathcal{U}} \mathcal{Z}_Z^*$.

Square root extraction problem : With the exponent fixed to $e = 2$, we call the problem of inverting $f_{pq}(y) = y^2 \bmod pq$ the square root extraction problem. For a given $z = pq$ and $y \in \mathcal{Z}_z^*$, $f_{pq}(y)$ has four inverses, viewed as two pairs $\langle y_0, z - y_0 \rangle$ and $\langle y_1, z - y_1 \rangle$, where both y_0 and y_1 are less than $z/2$. Given one member of either pair, the other member is trivial to compute given z. We describe how to find all four inverses of $f_{pq}(y)$ in $n^{\mathcal{O}(1)}$ time given the trapdoor key $\langle p, q \rangle$. This is done by computing the two square roots y_p, $p - y_p$ of $f_{pq}(y)$ with respect to p, computing the two square roots y_q, $q - y_q$ of $f_{pq}(y)$ with respect to q, and then combining all this information using the Chinese remainder algorithm to compute the four square roots $\langle y_0, z - y_0 \rangle$ and $\langle y_1, z - y_1 \rangle$ of y with respect to z.

---∞---

The following theorem shows that the square root extraction problem is as hard as the factoring problem (page 17).

Theorem 15.1 : The square root extraction problem is a trapdoor one-way function if the factoring problem is a one-way function. The reduction is linear-preserving.

PROOF: Let A be an adversary for inverting f with run time $T(n)$ Fix $z = pq$, let $Y^* \in_{\mathcal{U}} \mathcal{Z}_z^*$ and let

$$\delta_z = \Pr_{Y^*}[f_z(A(z, f_z(Y^*))) = f_z(Y^*)]$$

be the success probability of A with respect to z. Let $\langle\langle P,Q\rangle, Z\rangle \in_{\mathcal{D}_n}$ $\{0,1\}^{2n} \times \{0,1\}^{2n}$. The overall success probability of A is $\delta(n) = \mathrm{E}_Z[\delta_Z]$.

We describe an oracle machine S such that S^A factors Z with probability $\delta(n)/2$ and such that the running time of S^A is $\mathcal{O}(T(n))$. The input to S^A is $z = pq$,

---

Adversary S^A on input z : .

Choose $y \in_{\mathcal{U}} \mathcal{Z}_z$ and compute $f_z(y)$.

Compute $a = \gcd(z, f_z(y))$.

If $a \neq 1$ then output $\langle z/a, a\rangle = \langle p, q\rangle$ and stop.

Compute $y' = A(z, f_z(y))$.

Compute $b = \gcd(z, y + y')$.

Output $\langle z/b, b\rangle$. (This is equal to $\langle p, q\rangle$ if $b \neq 1$.)

Let $Y \in_{\mathcal{U}} \mathcal{Z}_z$. If $a \neq 1$ then S^A immediately factors z. The conditional distribution on Y given that $a = \gcd(z, f_z(Y)) = 1$ is Y^*, and we assume this for the remainder of the proof. Let $Y' = A(z, f_z(Y))$. Suppose that $f_z(Y') = f_z(Y)$, i.e., A is able to invert. Let $\langle Y_0, z - Y_0\rangle$ and $\langle Y_1, z - Y_1\rangle$ be the inverses of $f_z(Y)$ with respect to z, and without loss of generality suppose $Y' \in \{Y_0, z - Y_0\}$. Since the distribution on Y is uniform on the four inverses of $f_z(Y)$, with probability $1/2$ it is the case that $Y \in \{Y_1, z - Y_1\}$. Suppose $Y = Y_1$. Since $Y_0 + Y_1 \neq 0 \bmod z$ and since $Y_0 - Y_1 \neq 0 \bmod z$ and since $(Y_0 + Y_1)(Y_0 - Y_1) = Y_0^2 - Y_1^2 = 0 \bmod z$, it follows that $b = \gcd(Y_0 + Y_1, z) \in \{p, q\}$.

The overall probability S^A factors z is at least $\delta_z/2$. Thus, S^A factors Z with probability at least $\delta(n)/2$. ∎

---

one-way predicate

The notion of a one-way predicate is closely related to the notion of a one-way function. Intuitively, a one-way predicate f is a **P**-time function ensemble which, in addition to other inputs, has a $\{0,1\}$-valued input b that is hard to predict given the output of f, but nevertheless b is uniquely determined by the output of f.

Definition (one-way predicate): Let $f : \{0,1\}^{\ell(n)} \times \{0,1\} \times \{0,1\}^n \to \{0,1\}^{k(n)}$ be a **P**-time function ensemble with the additional property that, for all $z \in \{0,1\}^{\ell(n)}$, for all $y, y' \in \{0,1\}^n$,

$$f_{z,0}(y) \neq f_{z,1}(y),$$

i.e., with respect to fixed z and y, the value of $b \in \{0,1\}$ is uniquely determined by z and $f_{z,b}(y)$. The first input is public and the second and third are private, and thus the security parameter $\mathbf{s}(n)$ is $n+1$. Let $A : \{0,1\}^{\ell(n)} \times \{0,1\}^{k(n)} \to \{0,1\}$ be an adversary. Let $Z \in_{\mathcal{U}} \{0,1\}^{\ell(n)}$, $B \in_{\mathcal{U}} \{0,1\}$, and $Y \in_{\mathcal{U}} \{0,1\}^n$. The success probability of A for f is

$$\delta(n) = |\mathrm{E}[\overline{A(Z, f_{Z,B}(Y))} \cdot \overline{B}]|.$$

Then, f is a $\mathbf{S}(\mathbf{s}(n))$-*secure one-way predicate* if every adversary has time-success ratio at least $\mathbf{S}(\mathbf{s}(n))$. ♣

We can view the input bit b of f as a bit that is statistically committed but still hidden given the output value of f. (See the hidden bit commitment protocol on page 181.)

Construction of a one-way predicate : Let $f : \{0,1\}^n \to \{0,1\}^n$ be a one-way permutation. Let $z \in \{0,1\}^n$, $b \in \{0,1\}$, and $y \in \{0,1\}^n$. Define one-way predicate $g_{z,b}(y) = \langle f(y), b \oplus (y \odot z) \rangle$.

Exercise 58 : Prove that if f is a one-way permutation then g is a one-way predicate. The reduction should be linear-preserving.

Hint : See the Hidden Bit Theorem on page 65. ♠

A trapdoor one-way predicate is a one-way predicate with a trapdoor key that allows the $\{0,1\}$-valued input of the function to be easily computed given the output of the function.

Definition (trapdoor one-way predicate): Let $\mathcal{D}_n : \{0,1\}^{r(n)} \to \{0,1\}^{m(n)+\ell(n)}$ be a **P**-samplable probability ensemble. We call $\mathcal{D}_n$ the *key generation distribution*. Let $\langle x, z \rangle \in \{0,1\}^{m(n)} \times \{0,1\}^{\ell(n)}$ be a possible output of $\mathcal{D}_n$. We call x the *trapdoor key* and z the *public key*. Let $f : \{0,1\}^{\ell(n)} \times \{0,1\} \times \{0,1\}^n \to \{0,1\}^{k(n)}$ be a **P**-time function ensemble with the additional property that, for all $z \in \{0,1\}^{\ell(n)}$, for all $y, y' \in \{0,1\}^n$,

$$f_{z,0}(y) \neq f_{z,1}(y),$$

i.e., with respect to fixed z and y, the value of $b \in \{0,1\}$ is uniquely determined by z and $f_{z,b}(y)$. The first input is the public key and the second and third are private. The following properties hold:

(invertible with the trapdoor key) Let $y \in \{0,1\}^n$. There is a **P**-time function ensemble that on input $\langle z, f_{z,b}(y), x \rangle$ produces b.

(one-way predicate) Let $\langle X, Z \rangle \in_{\mathcal{D}_n} \{0,1\}^{m(n)} \times \{0,1\}^{\ell(n)}$, $B \in_{\mathcal{U}} \{0,1\}$, and $Y \in_{\mathcal{U}} \{0,1\}^n$. We view X, B, and Y as private, and thus the security parameter $\mathbf{s}(n)$ is $m(n)+n+1$. Let $A : \{0,1\}^{\ell(n)} \times \{0,1\}^{k(n)} \to \{0,1\}$ be an adversary. The success probability of A for f is

$$\delta(n) = |\mathrm{E}[\overline{A(Z, f_{Z,B}(Y))} \cdot \overline{B}]|.$$

Then, f is a $\mathbf{S}(\mathbf{s}(n))$-*secure trapdoor one-way predicate* if every adversary has time-success ratio at least $\mathbf{S}(\mathbf{s}(n))$.   ♣

The construction of a one-way predicate based on a one-way permutation given above also yields a trapdoor one-way predicate based on a trapdoor one-way permutation.

A specific problem related to the factoring and the square root extraction problems that is conjectured to be a trapdoor one-way predicate is the following.

Quadratic residuosity problem : Let $\langle p, q \rangle$ be a pair of n-bit primes, and let $z = pq$. Let $y \in \mathcal{Z}_z^*$. The Jacobi symbol $J_z(y)$ is a $\{1, -1\}$-valued **P**-time function ensemble. Let $\mathcal{J}_z = \{y \in \mathcal{Z}_z^* : J_z(y) = 1\}$. Let $\alpha \in \mathcal{Z}_z^*$ be a fixed non-square with Jacobi symbol 1, i.e., $y^2 \bmod z \neq \alpha$ for all $y \in \mathcal{Z}_z^*$ and $\alpha \in \mathcal{J}_z$. Let $\mathcal{Q}_z = \{y^2 \bmod z : y \in \mathcal{Z}_z^*\}$ be the set of squares mod z, and let $\bar{\mathcal{Q}}_z = \{\alpha y^2 \bmod z : y \in \mathcal{Z}_z^*\}$ be the set of non-squares mod z with Jacobi symbol 1. Define trapdoor one-way predicate $f_{z,0}(y) = y^2 \bmod z$ and $f_{z,1}(y) = \alpha y^2 \bmod z$. Note that it is possible in $n^{\mathcal{O}(1)}$ time to compute the value of b given $\langle p, q \rangle$, $f_{z,b}(y)$, and α. The key generation distribution on which this predicate is conjectured to be hard to predict is when the pair of primes $\langle P, Q \rangle$ is randomly chosen so that $Z = PQ$ is hard to factor on average.

Cryptosystems without initial private communication

For all previously described cryptosystems, there is an initialization stage where a private line is used to establish commonly shared private information. Thereafter, all communication is via a public line. The initial communication using a private line is sometimes infeasible to enact in certain physical situations, and thus it is desirable to have a cryptosystem that doesn't rely on a private line. Based on trapdoor one-way predicates, we describe a cryptosystem that achieves this. Suppose that party P_1 wants to send encrypted messages to party P_2. Let

$f : \{0,1\}^{\ell(n)} \times \{0,1\} \times \{0,1\}^n \to \{0,1\}^{k(n)}$ be a trapdoor one-way predicate with key generation distribution $\mathcal{D}_n : \{0,1\}^{r(n)} \to \{0,1\}^{m(n)+\ell(n)}$.

Public key bit cryptosystem :

- P_2 uses $r(n)$ random bits to produce $\langle x, z \rangle \in_{\mathcal{D}_n} \{0,1\}^{m(n)} \times \{0,1\}^{\ell(n)}$. P_2 sends the public key z to P_1 on a public line, and keeps the trapdoor key x private.

- Suppose that P_1 wants to send the message bit $b \in \{0,1\}$. P_1 chooses $y \in_{\mathcal{U}} \{0,1\}^n$ and sends $f_{z,b}(y)$ to P_2 on the public line.

- P_2 recovers b from $f_{z,b}(y)$, the public key z, and the trapdoor key x.

Exercise 59 : Prove that if f is a trapdoor one-way predicate then the above cryptosystem is secure against chosen plaintext attack. The reduction should be linear-preserving. ♠

This probabilistic encryption scheme has advantages and disadvantages compared to a stream cryptosystem. It is better because the encryption of each bit does not depend on an index. It is worse because the length of an encryption of each bit is long. The following construction of a block cryptosystem based on a trapdoor one-way permutation overcomes these problems.

Public key block cryptosystem :

(initialization): Let $f : \{0,1\}^{\ell(n)} \times \{0,1\}^n \to \{0,1\}^n$ be a trapdoor one-way permutation and let $\mathcal{D}_n$ be the key generation distribution on $\{0,1\}^{m(n)} \times \{0,1\}^{\ell(n)}$ associated with f. Suppose that P_2 wants to send messages to P_1. P_1 chooses $\langle x, z \rangle$ randomly according to $\mathcal{D}_n$, sends the public key z to P_2 on a public line, and keeps the trapdoor key x private.

(message sending): Suppose P_2 wants to send message block $m \in \{0,1\}^{p(n)}$ to P_1. P_2 randomly chooses $r \in_{\mathcal{U}} \{0,1\}^n$ and $y \in_{\mathcal{U}} \{0,1\}^n$. Then, for all $i \in \{1, \ldots, p(n)\}$, P_2 computes

$$b_i = (f_z^{(i-1)}(y) \odot r) \oplus m_i,$$

where $f_z^{(0)}(y) = y$ and for $i \geq 1$, $f_z^{(i)}(y) = f_z(f_z^{(i-1)}(y))$. Let $b = \langle b_1, \ldots, b_{p(n)} \rangle$. P_2 sends to P_1 the encryption $\langle b, r, f_z^{(p(n))}(y) \rangle$ on a public line. Upon receiving this, P_1 can decrypt as follows. Since P_1 knows the trapdoor key x, P_1 can compute

$$\langle f_z^{(0)}(y), \ldots, f_z^{(p(n)-1)}(y) \rangle$$

from $f_z^{(p(n))}(y)$. Then, for all $i \in \{1, \ldots, p(n)\}$, P_1 can compute

$$m_i = (f_z^{(i-1)}(y) \odot r) \oplus b_i.$$

Exercise 60 :   Prove that if f is a trapdoor one-way permutation then the the block cryptosystem just described is secure against chosen plaintext attack. The proof should be linear-preserving.   ♠

Exchanging Secret Keys

Suppose a party P_1 has privately chosen a pair of primes $\langle p, q \rangle$, and P_2 wants to send $\langle p, q \rangle$ to another party P_1 on a public line without leaking this information to an adversary. The next protocol shows how this can be achieved, assuming that factoring $z = pq$ is hard.

Secret Factorization Exchange Protocol :

- P_1 sends $z = pq$ to P_2.

- P_2 chooses $y \in_\mathcal{U} \mathcal{Z}_z^*$ and sends $x = y^2 \bmod z$ to P_1.

- P_1 computes the four square roots $\langle y_0, z - y_0 \rangle$ and $\langle y_1, z - y_1 \rangle$ of x with respect to z, selects $y' \in_\mathcal{U} \{y_0, z - y_0, y_1, z - y_1\}$ and sends y' to P_2.

- If $y \neq y'$ and $y \neq z - y'$ then P_2 computes $\gcd(y + y', z)$ and from this obtains the factors p and q of z.

With probability $1/2$ the square root y' that P_1 sends to P_2 is not of the form $y' = y$ or $y' = z - y$, and thus as explained in the proof of Theorem 15.1, $\gcd(y + y', x) \neq 1$. Thus, with probability $1/2$, P_2 factors z. Moreover, an adversary A has no extra information about the factorization of z from the conversation (except that A knows z of course). To see this, observe that A can simulate the entire conversation after the first step. A can choose y randomly just as P_2 does and simulate sending $y^2 \bmod z$ to P_1, and then A can simulate P_1 by sending y back to P_2. It is not hard to show that the distribution on messages as seen by A in the actual conversation is exactly the same as in this simulation and hence that this is a faithful simulation of the conversation. From this it follows that A has no advantage factoring z seeing the conversation then A has without seeing the conversation.

After enacting the above protocol, P_2 fails to know the factorization with probability $1/2$. This failure probability can be decreased to $2^{-\ell}$ by running the last three steps independently ℓ times. If $p = q = 3 \bmod 4$, then the protocol can be modified so that P_2 always receives the factorization after one round. In this case it turns out that the square roots $\langle y_0, z - y_0 \rangle$ and $\langle y_1, z - y_1 \rangle$ of x with respect to z have the following property with respect to the Jacobi symbol:

$$J_z(y_0) = J_z(z - y_0) \neq J_z(y_1) = J_z(z - y_1).$$

The protocol is modified so that in the second step of the protocol, P_2 sends $\langle x, J_z(y) \rangle$ to P_1, and then in the third step P_1 sends square root y' of x to P_2 where $J_z(y') \neq J_z(y)$. Based on the proof of Theorem 15.1, P_2 can factor z by computing $\gcd(y + y', z)$ in the fourth step.

Lecture 16

Overview

We give the definition and a construction of a universal one-way hash function. One of the main technical tools we use to construct a secure digital signature scheme in the next lecture is a universal one-way hash function. A universal one-way hash function is also interesting in its own right.

Definition of a universal one-way hash function

Intuitively, a universal one-way hash function is like a universal hash function (page 84) with security properties. As described in the next lecture, a universal one-way hash function is a useful tool in the construction of secure digital signature schemes.

Definition (universal one-way hash function): Let $g : \{0,1\}^n \times \{0,1\}^{d(n)} \rightarrow \{0,1\}^{r(n)}$ be a **P**-time function ensemble, where $r(n) < d(n)$. For a fixed $y \in \{0,1\}^n$, we view $g_y(x) = g(y,x)$ as a function of x from $\{0,1\}^{d(n)}$ to $\{0,1\}^{r(n)}$. The quantity $d(n) - r(n)$ is called the compression value of g. We let the security parameter $s(n) = r(n)$. Let A be an adversary that works as follows.

(Stage 1) Run A to produce a string $x \in \{0,1\}^{d(n)}$.

(Stage 2) Choose $y \in_{\mathcal{U}} \{0,1\}^n$ and give y to A.

(Stage 3) A tries to produce a string $x' \in \{0,1\}^{d(n)} \setminus \{x\}$ such that $g_y(x') = g_y(x)$.

The success probability $\delta(n)$ of A is the probability, with respect to randomly chosen $y \in_{\mathcal{U}} \{0,1\}^n$, that A in Stage 3 produces $x' \neq x$ such that $g_y(x') = g_y(x)$. Then, g is a $\mathbf{S}(s(n))$-*secure universal one-way hash function* if every adversary has time-success ratio at least $\mathbf{S}(s(n))$. ♣

In our applications of a universal one-way hash function, a party produces $y \in_{\mathcal{U}} \{0,1\}^n$ and $x_1, \ldots, x_{p(n)} \in_{\mathcal{U}} \{0,1\}^{d(n)}$ independently distributed, where $p(n)$ is a polynomial parameter. The adversary is only involved in Stage 3, i.e., the adversary receives all of this and tries to find for some $i \in \{1, \ldots, p(n)\}$ an $x \neq x_i$ such that $g_y(x) = g_y(x_i)$. This motivates an alternative definition of a universal one-way hash function that is strong enough for our applications (and more directly applicable) but weaker than the definition given above.

Definition (alternative definition of a universal one-way hash function): An adversary A for a universal one-way hash function g : $\{0,1\}^n \times \{0,1\}^{d(n)} \to \{0,1\}^{r(n)}$ works as follows.

(Stage 1) Let $p(n)$ be a polynomial parameter. Party P chooses $y \in_\mathcal{U} \{0,1\}^n$ and $x_1, \ldots, x_{p(n)} \in_\mathcal{U} \{0,1\}^{d(n)}$ independently and gives this to the adversary A.

(Stage 2) A tries to produce a string $x \in \{0,1\}^{d(n)}$ such that for some $i \in \{1, \ldots, p(n)\}$, $x' \neq x_i$ but $g_y(x) = g_y(x_i)$.

The success probability $\delta(n)$ of A is the probability, with respect to $y \in_\mathcal{U} \{0,1\}^n$ and $x_1, \ldots, x_{p(n)} \in_\mathcal{U} \{0,1\}^{d(n)}$, that A in Stage 2 produces $x \neq x_i$ such that $g_y(x) = g_y(x_i)$. The security parameter is $s(n) = r(n)$. Then, g is a $\mathbf{S}(\mathbf{s}(n))$-*secure universal one-way hash function* if every adversary has time-success ratio at least $\mathbf{S}(\mathbf{s}(n))$. ♣

Exercise 61 : Show that if g is a universal one-way hash function with respect to the original definition then g is a universal one-way hash function with respect to the alternative definition. ♠

Research Problem 5 : Is there a universal one-way hash function with respect to the definition where the adversary A first sees $y \in_\mathcal{U} \{0,1\}^n$ and then tries to produce a pair $x, x' \in \{0,1\}^{d(n)}$, such that $x \neq x'$ but $g_y(x) = g_y(x')$? ♠

A hash function with special properties

The construction of a universal one-way hash function g that compresses by one bit consists of the composition of a hash function h with a one-way permutation f.

Definition (the domain of indices): The *domain of function indices* for h is

$$D_n = \{y = \langle y_1, y_2 \rangle \in \{0,1\}^{2 \times n} : y_1 \neq 0^n\}.$$

♣

The construction of h is based on the Linear Polynomial Space (page 57).

Construction of h : Let $h : \{0,1\}^{2n} \times \{0,1\}^n \to \{0,1\}^{n-1}$ be a P-time function ensemble with compression value 1 that is defined as follows.

Let $y = \langle y_1, y_2 \rangle \in D_n$ and $x \in \{0,1\}^n$. Define

$$h'_y(x) = y_1 \cdot x + y_2,$$

where, on the right-hand side of the equal sign, y_1, y_2 and x are viewed as elements of $GF[2^n]$ and the field operations are with respect to $GF[2^n]$, and the result is viewed as an element of $\{0,1\}^n$. Then, for all $y \in D_n$ and for all $x \in \{0,1\}^n$ we define

$$h_y(x) = h'_y(x)_{\{1,\ldots,n-1\}},$$

i.e., $h_y(x)$ is obtained from $h'_y(x)$ by chopping off the last bit. This particular hash function $h_y(x)$ has properties listed below that are useful in the construction of a universal one-way hash function.

Properties of h :

- For each $y = \langle y_1, y_2 \rangle \in D_n$ and for each $z \in \{0,1\}^{n-1}$,

$$\sharp \mathrm{pre}_{h_y}(z) = 2,$$

  i.e., h_y is a two-to-one onto function for all $y \in D_n$. This is because, for each $y \in D_n$, h'_y is a permutation and pairs of elements in the range of h'_y are mapped to the same string by h_y.

- Fix $x \in \{0,1\}^n$ and $x' \in \{0,1\}^n \setminus \{x\}$. Define

$$D(x, x') = \{y \in D_n : h_y(x) = h_y(x')\}.$$

  Let $Z \in_{\mathcal{U}} \{0,1\}^n$. There is a **P**-time function ensemble M : $\{0,1\}^n \times \{0,1\}^n \times \{0,1\}^n \to \{0,1\}^{2n}$ such that $M(x, x', Z) \in_{\mathcal{U}} D(x, x')$. To see this, we first describe $D(x, x')$. Since $y = \langle y_1, y_2 \rangle \in D_n$ implies that $y_1 \neq 0^n$, any $y \in D_n$ that satisfies $h_y(x) = h_y(x')$ equivalently satisfies

$$y_1 \cdot x + y_2 = y_1 \cdot x' + y_2 + 1$$

  over $GF[2^n]$ (Note that $1 + 1 = 0$ over $GF[2^n]$, and thus it doesn't matter on which side of the equality the $+1$ is written.) Equality in this equation is independent of y_2, and holds if and only if $y_1 = (x - x')^{-1}$ over $GF[2^n]$. Thus, $M(x, x', Z)$ first computes $y_1 = (x - x')^{-1}$ over $GF[2^n]$ and produces $\langle y_1, Z \rangle$.

- Fix $x \in \{0,1\}^n$, and let $X' \in_{\mathcal{U}} \{0,1\}^n \setminus \{x\}$. Then, $M(x, X', Z) \in_{\mathcal{U}} D_n$. This is because $M(x, X', Z) = \langle (x - X')^{-1}, Z \rangle$, and because $(x - X')^{-1} \in_{\mathcal{U}} GF[2^n] \setminus \{0\}$.

The second property listed above is the "collision accessible" property of the hash function.

Compressing by one bit

Construction with compression value one : Let $h : \{0,1\}^{2n} \times \{0,1\}^n \rightarrow \{0,1\}^{n-1}$ be the hash function described in the previous section and let $f : \{0,1\}^n \rightarrow \{0,1\}^n$ be a one-way permutation. Let $g : \{0,1\}^{2n} \times \{0,1\}^n \rightarrow \{0,1\}^{n-1}$ be a **P**-time function ensemble defined as follows. The domain of function indices for g is the same as for h, i.e., the domain is D_n. For all $y \in D_n$ and for all $x \in \{0,1\}^n$, define

$$g_y(x) = h_y(f(x)).$$

The security parameter of g is n.

Theorem 16.1 : If f is a one-way permutation then g is a universal one-way hash function. The reduction is linear-preserving.

PROOF: Suppose there is an adversary A for g with success probability $\delta(n)$. We describe an oracle adversary S such that S^A is an adversary for f. The input to S^A is $f(w)$ where $w \in \{0,1\}^n$.

Adversary S^A on input $f(w)$: .

Run Stage 1 of A to produce a string $x \in \{0,1\}^n$.

Choose $z \in_{\mathcal{U}} \{0,1\}^n$, let $y = M(f(w), f(x), z)$ and give y to A.

Run Stage 3 of A to produce a string $x' \in \{0,1\}^n$.

Output x'.

Let $W \in_{\mathcal{U}} \{0,1\}^n$. We prove that, with probability at least $\delta(n) - 2^{-n}$, the output of S^A is W when the input to S^A is $f(W)$. Because f is a permutation, $f(W) \in_{\mathcal{U}} \{0,1\}^n$. Furthermore, W is independent of the x that A produces in Stage 1, and thus $f(W)$ is independent of $f(x)$. $f(W)$ is equal to $f(x)$ with probability 2^{-n}. Let $Y = M(f(W), f(x), Z)$. Because of the properties of h (page 156), $Y \in_{\mathcal{U}} D_n$ conditional on $f(W) \neq f(x)$. With respect to this conditional distribution, A produces x' such that $x \neq x'$ and

$$h_Y(f(x)) = g_Y(x) = g_Y(x') = h_Y(f(x'))$$

with probability $\delta(n)$. Because f is a permutation, $x \neq x'$ implies $f(x) \neq f(x')$. By the properties of M,

$$g_Y(W) = h_Y(f(W)) = h_Y(f(x)) = g_Y(x).$$

Because h_Y is a two-to-one function, if $f(W) \neq f(x)$ and $f(x) \neq f(x')$ then it must be the case that $f(W) = f(x')$, and from this it follows that $x' = W$. Overall, the probability that this event occurs is at least $\delta(n) - 2^{-n}$.   ∎

Compressing by many bits

We construct a universal one-way hash function that compresses by many bits by using several compositions of a universal one-way hash function that compresses by a single bit. We want the universal one-way hash function that compresses by many bits to be based on the difficulty of inverting a one-way permutation $f : \{0,1\}^n \rightarrow \{0,1\}^n$ for a fixed value of n independent of the number of bits to be compressed. This is desirable because in practice we may only have a one-way permutation for a specific value of n, i.e., not for all values of n. To do this, we first introduce a slight variant of the universal one-way hash function that compresses by one bit.

Construction for compressing a large number of bits by one : Let $t(n)$ be a polynomial parameter. Let

$$h : \{0,1\}^{2(n+t(n))} \times \{0,1\}^{n+t(n)} \rightarrow \{0,1\}^{n+t(n)-1}$$

be the hash function previously described and let $f : \{0,1\}^n \rightarrow \{0,1\}^n$ be a one-way permutation. Let

$$g^{t(n)} : \{0,1\}^{2(n+t(n))} \times \{0,1\}^{n+t(n)} \rightarrow \{0,1\}^{n+t(n)-1}$$

be a **P**-time function ensemble defined as follows. For all $y \in D_{n+t(n)}$, for all $x \in \{0,1\}^n$, for all $r \in \{0,1\}^{t(n)}$, define

$$g_y^{t(n)}(\langle x, r \rangle) = h_y(\langle f(x), r \rangle).$$

Exercise 62 : Show that if f is a one-way permutation then $g^{t(n)}$ is a universal one-way hash function. The reduction should be linear-preserving.

Hint :   Let $f^{t(n)}(\langle x, r \rangle) = \langle f(x), r \rangle$. Show that $f^{t(n)}$ is a one-way permutation.   ♠

Construction with large compression value : Let $t(n)$ be a polynomial parameter. Let $g : \{0,1\}^{\ell(n)} \times \{0,1\}^{n+t(n)} \to \{0,1\}^n$ be the **P**-time function ensemble defined as follows, where $\ell(n) = t(n)(2n + t(n) + 1)$. Let $y = \langle y_1, \ldots, y_{t(n)} \rangle \in \{0,1\}^{\ell(n)}$, where, for all $i \in \{1, \ldots, t(n)\}$, $y_i \in D_{n+i}$. Let $x \in \{0,1\}^n$ and $r \in \{0,1\}^{t(n)}$. Define

$$g_y(\langle x, r \rangle) = g^1_{y_1}(g^2_{y_2}(\cdots g^{t(n)}_{y_{t(n)}}(\langle x, r \rangle)) \cdots),$$

where g^i is as defined in the previous construction. The security parameter of this construction is n.

Theorem 16.2 : If f is a one-way permutation then g is a universal one-way hash function. The reduction is linear-preserving.

PROOF: Let A be an adversary for g with run time $T(n)$ and success probability $\delta(n)$. A first produces $x \in \{0,1\}^n$ and and $r \in \{0,1\}^{t(n)}$. Then, A receives a random $y = \langle y_1, \ldots, y_{t(n)} \rangle$, and produces an $x' \in \{0,1\}^n$ and $r' \in \{0,1\}^{t(n)}$ with $\langle x, r \rangle \neq \langle x', r' \rangle$ and $g_y(\langle x, r \rangle) = g_y(\langle x', r' \rangle)$ with probability at least $\delta(n)$. We describe an oracle adversary S such that S^A has success probability $\delta'(n) = \delta(n)/t(n)$ for the universal one-way hash function g^i that compresses by one bit described in the previous construction, where $i \in_U \{1, \ldots, t(n)\}$. Then, the proof is completed by appealing to Exercise 62, which is a generalization of Theorem 16.1.

For all $i \in \{1, \ldots, t(n)\}$, for all $j \in \{i, \ldots, t(n)\}$, for all $x \in \{0,1\}^n$, and for all $r \in \{0,1\}^j$, for all $y_k \in D_{n+k}$ for $k \in \{i, \ldots, j\}$, define

$$g_{y_{\{i,\ldots,j\}}}(\langle x, r \rangle) = g^i_{y_i}(g^{i+1}_{y_{i+1}}(\cdots (g^j_{y_j}(\langle x, r \rangle)) \cdots).$$

For consistency in notation, define

$$g_{y_{\{j+1,j\}}}(\langle x, r \rangle) = \langle x, r \rangle.$$

Note that if, for some $\langle x, r \rangle \in \{0,1\}^{n+t(n)}$ and $\langle x', r \rangle \in \{0,1\}^{n+t(n)}$, both of the following are true:

- $\langle x, r \rangle \neq \langle x', r' \rangle$

- $g_y(\langle x, r \rangle) = g_y(\langle x', r' \rangle)$

then there must be some $i \in \{1, \ldots, t(n)\}$ such that the following conditions hold:

- $g_{y_{\{i+1,\ldots,t(n)\}}}(\langle x, r \rangle) \neq g_{y_{\{i+1,\ldots,t(n)\}}}(\langle x', r' \rangle)$

- $g_{y_{\{i,\ldots,t(n)\}}}(\langle x, r \rangle) = g_{y_{\{i,\ldots,t(n)\}}}(\langle x', r' \rangle)$

For a fixed value of i, let $\delta_i(n)$ be the probability that these conditions hold with respect to adversary A. Since

$$\sum_{i \in \{1,\ldots,t(n)\}} \delta_i(n) = \delta(n),$$

$$E_{I \in_{\mathcal{U}} \{1,\ldots,t(n)\}}[\delta_I(n)] \geq \frac{\delta(n)}{t(n)}.$$

S^A chooses $i \in_{\mathcal{U}} \{1,\ldots,t(n)\}$, and then tries to break g^i as a universal one-way hash function as follows: The first stage of S^A uses the first stage of A in its attack on g to produce $x \in \{0,1\}^n$ and $r \in \{0,1\}^{t(n)}$. Then, in the second stage of the attack of A on g, A is given $y = \langle y_1, \ldots, y_{t(n)} \rangle$ independently and uniformly chosen. The output of S^A in the first stage of its attack on g^i is then

$$\langle x_i, r_i \rangle = g_{y_{\{i+1,\ldots,t(n)\}}}(\langle x, r \rangle),$$

and $y_1, \ldots, y_i$ is the output of the second stage of the attack of S^A on g^i, In the third stage of the attack of S^A on g^i, the third stage of the attack of A on g is run to produce $\langle x', r' \rangle \in \{0,1\}^{n+t(n)}$. Finally S^A produces

$$\langle x'_i, r'_i \rangle = g_{y_{\{i+1,\ldots,t(n)\}}}(\langle x', r' \rangle).$$

Since y is random and independent of x,

$$\langle x_i, r_i \rangle \neq \langle x'_i, r'_i \rangle$$

and

$$g^i_{y_i}(\langle x_i, r_i \rangle) = g^i_{y_i}(\langle x'_i, r'_i \rangle)$$

both hold with probability $\delta_i(n)$. Since $i \in_{\mathcal{U}} \{1,\ldots,t(n)\}$, the success probability of S^A is $\delta(n)/t(n)$. ∎

Suppose $m(n) \gg n$ and consider designing a universal one-way hash function g that maps $\{0,1\}^{m(n)}$ down to $\{0,1\}^n$. In this case, the length of the description of the hash function in Theorem 16.2 is fairly long in terms of the final output of the hash function. The following exercise shows how to get around this problem.

Exercise 63 :   Design a universal one-way hash function $h_y(x)$ that maps $x \in \{0,1\}^{m(n)}$ down to $h_y(x) \in \{0,1\}^n$ such that the length of y is

$$\mathcal{O}(n^2 \cdot \log(m(n)/n)).$$

For example, if $m(n)$ is set to n^3, then the universal one-way hash function should map n^3 bits down to n bits using a hash function description of length $\mathcal{O}(n^2 \log(n))$.

Hint : Break the original input up into $m(n)/n$ blocks, each of length n, and use one universal one-way hash function that maps n bits downs to $n/2$ bits to simultaneously map all $m(n)/n$ blocks down to length $n/2$ each. Apply the same technique $\log(m(n))/n$ times, using independently chosen hash functions each time. ♠

Lecture 17

Overview

We give the definition and the construction of a signature scheme based on a universal one-way hash function.

Signing One Message

A one message signature scheme is a way for a party S (called the signer) to create a signature σ of a message m, and send the pair $\langle m, \sigma \rangle$ to another party V (called the verifier). Intuitively, the scheme is secure if S is the only party that can convince V that S signed m, even in the case when V cannot be sure whether it is an adversary or S that sends $\langle m, \sigma \rangle$.

There are three phases to the scheme, an initialization phase, a signature phase and a verification phase. In the initialization phase, V and S are allowed to use a public line. In this phase, S performs some computation to produce a key $\langle s, v \rangle$, where s is called the private part of the key and is kept private by S, and v is called the public part of the key and is given to V. An important point is that because V and S are using a public line, V has a guarantee that v actually came from S. However, there is no assumption about other parties not being able to see the exchange of information between S and V, and in particular a potential adversary A is assumed to know the public part v of the key.

After the initialization phase, S and V are only allowed to communicate via a public network. At some later point in time, S wants to send a message m to V and convince V that S generated the message. S uses the signature phase to do this: S computes the information σ based on $\langle s, v, m \rangle$, where σ is called the signature of m. S sends $\langle m, \sigma \rangle$ to V. When S sends this information to V, since S and V are using a public network, V has no guarantee that what is received really came from S.

In the final phase, the verification phase, V does some computation based on $\langle v, m, \sigma \rangle$ received in the previous two phases and decides whether or not to believe that it was S who sent m to V in the signature phase. What V is trying to protect against is an adversary A who wants to send a message m to V and convince V that it was actually S who sent m to V. What A may try to do is forge a signature σ' of m and send $\langle m, \sigma' \rangle$ to V in the signature phase. The phases are designed in a way that protects

V from this type of attack. Intuitively, the protection is based on the following. Because S and V enact the initialization phase using a public line, V knows that v was sent by S. Part of the verification procedure enacted by V is to check that $\langle m, \sigma' \rangle$ is a valid signature with respect to v. It turns out that s and v are connected in a crucial way, i.e., a valid signature σ of m with respect to v can be easily computed from the key $\langle s, v \rangle$, but it is hard to generate a valid signature of m knowing only v. Since A doesn't know s, this makes it hard for A to forge a valid signature of m.

Protection is also provided to S in the following sense. Suppose all interested parties are gathered together in the initialization phase, including all possible adversaries, and they all verify that S is the one who sends the public information in this phase. If the scheme is secure then S is protected against forged signatures in the sense that nobody, including the verifier V, can forge a signature of S on a message and convince anyone else that S signed the message.

Definition (one message scheme): A *one message signature scheme* is a pair $\langle S, V \rangle$, where S and V are randomized **P**-time function ensembles that interact as follows.

(initialization): S creates the key $\langle s, v \rangle$ privately, sends v to V on a public line and stores s in private memory.

(signature): S uses $\langle s, v, m \rangle$ to produce the signature σ of a message m. S sends $\langle m, \sigma \rangle$ to V on public network.

(verification): Upon receiving $\langle m, \sigma \rangle$, V performs some computation based on $\langle v, m, \sigma \rangle$ and either accepts or rejects.

The security parameter $s(n)$ of the scheme is the total length of all the private information that S keeps for the duration of the scheme. ♣

We define two different types of security for a one message scheme. In the first definition of security, an adversary A is allowed to try to sign a single message chosen at random from a distribution, and the adversary is successful if the verifier accepts the forged signature produced by A.

Definition (distributionally secure): Let $\mathcal{D}_n$ be **P**-samplable distribution and let $\langle S, V \rangle$ be a one message signature scheme.

(initialization): S creates $\langle s, v \rangle$ and sends v to V on a public line (A also receives v).

(attempted forged signature): Choose $m \in_{\mathcal{D}_n} \{0,1\}^n$ and give m to A. Using $\langle v, m \rangle$, A produces σ' and sends $\langle m, \sigma' \rangle$ to V on a public network.

(verification): Upon receiving $\langle m, \sigma' \rangle$, V performs some computation based on $\langle v, m, \sigma' \rangle$ and either accepts or rejects.

The success probability $\delta(n)$ of A is defined as the probability that V accepts in the verification phase, where the probability is over the output of S in the initialization phase, the choice of m and the random choices of the verifier V. We say that $\langle S, V \rangle$ is $\mathbf{S}(\mathbf{s}(n))$-*secure distributionally* with respect to $\mathcal{D}_n$ if every adversary has time-success ratio at least $\mathbf{S}(\mathbf{s}(n))$. ♣

Note that m is not chosen by A but given to A. This models the situation where the message A wants to forge a signature for is determined by some random outside force outside of the control of A, e.g., the outcome of a sporting event.

Unfortunately, in some situations distributional security is inadequate even in the situation where there is only one message to sign. In particular, a distributionally secure message scheme is not necessarily secure against an adversary who wants to forge the signature of an arbitrary message. In the following stronger definition, it is the adversary who chooses the message to be signed.

Definition (worst case secure):

(initialization): S creates $\langle s, v \rangle$ and sends v to V on a public line (A also receives v).

(attempted forged signature): A produces $\langle m, \sigma' \rangle$ based on v and sends $\langle m, \sigma' \rangle$ to V on a public network.

(verification): Upon receiving $\langle m, \sigma' \rangle$, V performs some computation based on $\langle v, m, \sigma' \rangle$ and either accepts or rejects.

The success probability $\delta(n)$ of A is defined as the probability that V accepts in the verification phase, where the probability is over the output of S in the initialization phase and the random choices of the verifier V. We say that $\langle S, V \rangle$ is $\mathbf{S}(\mathbf{s}(n))$-*secure in the worst case* if every adversary has time-success ratio at least $\mathbf{S}(\mathbf{s}(n))$. ♣

Square root signature scheme :

(**initialization**): S chooses at random two n-bit primes p and q. The key is $\langle s, v \rangle$, where $s = \langle p, q \rangle$ and $v = pq$. S sends v to V on a public line.

(**signature**): Assume that the message that S wants to sign is $m \in \mathcal{Q}_v$. S computes the four square roots σ_1, σ_2, σ_3, and σ_4 of $m \bmod v$, chooses $i \in_{\mathcal{U}} \{1, \ldots, 4\}$ and sends $\langle m, \sigma_i \rangle$ to V on a public network.

(**verification**): Upon receiving $\langle m, \sigma_i \rangle$, V checks to see if $\sigma_i^2 \bmod v = m$ and accepts if this is an equality and rejects otherwise.

S is able to compute the four square roots of $m \bmod v$ in the signature phase because S has the factors $\langle p, q \rangle$ of v. (See the definition of the Square root extraction problem on page 147). The security parameter $s(n)$ of this scheme is $\| s \| = 2n$.

Exercise 64 : Show that the Square root signature scheme is not worst case secure. ♠

Exercise 65 : Show that if factoring is hard then the Square root signature scheme is distributionally secure with respect to the message distribution that is uniform on $\mathcal{Q}_v$. The reduction should be linear-preserving.

Hint : Look at the proof of Theorem 15.1 (page 147). ♠

The following one bit scheme is worst case secure. This scheme is the starting point for building the signature scheme for signing multiple messages described in the next section.

One bit signature scheme : Let $f(x)$ be a one-way function.

(**initialization**): S chooses $x \in_{\mathcal{U}} \{0, 1\}^n$, and $y \in_{\mathcal{U}} \{0, 1\}^n$. S creates a window, which consists of two parts. The private part of the window is $s = \langle x, y \rangle$ and S computes the public part of the window as $v = \langle f(x), f(y) \rangle$. S sends v to V on a public line.

(**signature**): Let $b \in \{0, 1\}$ be the message that S wants to sign. Let $w = \langle s, v \rangle$ be the window created by S in the initialization phase. Let $s = \langle x, y \rangle$.

$$\text{if } \begin{cases} b = 0 & \text{then} \quad \sigma = x \\ b = 1 & \text{then} \quad \sigma = y \end{cases}.$$

S sends $\langle b, \sigma \rangle$ to V on a public network.

(**verification**): Let $v = \langle x', y' \rangle$. Upon receiving $\langle b, \sigma \rangle$,

$$\text{if } \begin{cases} b = 0 & \text{then} \quad V \text{ checks that } f(\sigma) = x' \\ b = 1 & \text{then} \quad V \text{ checks that } f(\sigma) = y' \end{cases} .$$

If the check yields equality then V accepts, else V rejects.

The security parameter $\mathbf{s}(n)$ of this scheme is $\| x \| + \| y \| = 2n$.

Exercise 66 :   Prove that if f is a one-way function then the One bit signature scheme is worst case secure. The reduction should be linear-preserving.   ♠

Signing Many Messages

Based on a one-way permutation, we describe a many message signature scheme. A one bit signature scheme can be easily modified to sign a many bit message by creating enough windows in the initialization phase to sign all the bits. However, in many scenarios it is unrealistic to assume that S and V know the total number of bits that are to be signed at this point. The scheme described below has the property that the number of messages that can be sent after the initialization phase is not limited, although of course security degrades with the number of messages sent.

Definition (many messages scheme):   A *many message signature scheme* is a pair $\langle S, V \rangle$, where S and V are randomized **P**-time function ensembles that interact as follows. In the initialization phase, S computes information and sends some portion of this information to V on a public line. When S wants to sign a new message, S computes information based on all of its previous computations and sends some portion of this information to V on a public network. When V wants to verify the signature of a message, V can use all information sent by S up to that point in time.   ♣

We need a stronger notion of security for a many messages scheme, i.e., the adversary A should not be able to forge the signature of any new message even after A interactively chooses messages and has S sign them.

Definition (security against adaptive chosen message attack): Let A be an adversary that is trying to forge a signature.

- S and V run the initialization phase using the public line. A sees all information sent on the line.

- A decides on message m_1 and S signs m_1.

- A decides on message m_2 and S signs m_2.

$$\vdots$$

- A decides on message m_i and S signs m_i.

In the initialization phase, the information sent by S is received by both A and V. After the initialization phase, all information sent by S is only received by A and not by V. Based on all the information received by A from S, A chooses a message $m \notin \{m_1, \ldots, m_i\}$ and interacts with V in the signature and verification phases. What A is trying to accomplish is to get V to accept m as a message that was signed by S. The success probability $\delta(n)$ of A is the probability that V is convinced that S signed m. The security parameter $\mathbf{s}(n)$ is the total length of all the private information that S must keep for the duration of the scheme. We say that $\langle S, V \rangle$ is $\mathbf{S}(\mathbf{s}(n))$-*secure against adaptive chosen message attack* if every adversary has time-success ratio at least $\mathbf{S}(\mathbf{s}(n))$. ♣

Adaptive chosen message security models the situation where A can get S to sign any sequence of messages except for some crucial message m for which A really wants to forge a signature. For example, S may be perfectly willing to sign messages like "It is probably going to rain tomorrow" and "S agrees to pay V \$5,050 next month if V gives S \$5,000 on the first day of this month" without caring too much about whether it was V or some other party to which S sends the message, signature pair, but for obvious reasons S probably would not agree to sign the message "S promises to pay A \$1,000,000 on the first of each month for the next two years, starting January 1."

A many message scheme that is secure against adaptive chosen message attack can be used as a one message scheme that is worst case secure.

A many message signature scheme : Assume all messages are n bits long. The idea is to use a block to sign each message of n bits. The blocks are linked together into a singly linked list, where the ith block in the list is used to sign the ith message. Define a component to be a sequence of n windows. See the One bit signature scheme (page 165) for a discussion of windows. Each block consists of two parts:

- A pointer component.

- A data component.

A key idea behind the construction is how make the link from one block to the next, which can be described as follows:

- A universal one-way hash function is used to compress the public parts of the next block to a short string,

- The pointer component is used to point to (commit to) the short string.

- The data component is used to sign (commit to) the ith message.

————∞————

Let $f : \{0,1\}^n \to \{0,1\}^n$ be a one-way permutation. The security of the scheme described below is based on the security of f. The reduction we describe from f to a secure multiple message signature scheme is only weak-preserving. This is because the total length of information that S keeps private is the total number of messages signed multiplied by $4n^2$, whereas a breaking adversary A for the scheme is converted into a breaking adversary M^A for f on inputs of length n.

A conceptually easier, but impossible to implement, method for linking one block to the next would be to use the pointer component of a given block to commit to the public information in the next block. The problem is that the pointer component can only commit to n bits, whereas the description of the public part of the next block is $4n^2$ bits. This is the reason we use the more complicated two part construction of a link using a universal one-way hash function as described above.

Let $g : \{0,1\}^{\ell(n)} \times \{0,1\}^{4n^2} \to \{0,1\}^n$ be a universal one-way hash function constructed using one-way permutation f as described in Lecture 16 on page 159, where $\ell(n) \approx 16n^4$. We need the following subprotocols to describe how the data and pointer components are used.

Creating a component : The following describes how S creates a component c. S chooses $x \in_\mathcal{U} \{0,1\}^{n \times n}$ and $y \in_\mathcal{U} \{0,1\}^{n \times n}$. The private part of the component is

$$s = \langle \langle x_1, y_1 \rangle, \ldots, \langle x_n, y_n \rangle \rangle$$

and S computes the public part of the component as

$$v = \langle \langle f(x_1), f(y_1) \rangle, \ldots, \langle f(x_n), f(y_n) \rangle \rangle.$$

The entire component is $c = \langle s, v \rangle$. We refer to this process as "S creates a component c".

Committing a component : Let $c = \langle s, v \rangle$ be a component originally created by S. The following describes how S can commit to a string $a \in \{0,1\}^n$ using c. Let

$$s = \langle \langle x_1, y_1 \rangle, \ldots, \langle x_n, y_n \rangle \rangle.$$

For all $i \in \{1, \ldots, n\}$,

$$\text{if} \begin{cases} a_i = 0 & \text{then} \quad \text{let } \alpha_i = x_i \\ a_i = 1 & \text{then} \quad \text{let } \alpha_i = y_i \end{cases}$$

Let $\alpha = \langle \alpha_1, \ldots, \alpha_n \rangle$. We refer to this process as "S uses c to commit to a" and we refer to α as the commitment of c to a.

Verifying a commitment : Let $c = \langle s, v \rangle$ be a component originally created by S, and suppose S has already used c to commit to $a \in \{0,1\}^n$ and $\alpha \in \{0,1\}^{n \times n}$ is the commitment of c to a. Then, in addition to a, V has

$$v = \langle \langle x_1', y_1' \rangle, \ldots, \langle x_n', y_n' \rangle \rangle$$

and

$$\alpha = \langle \alpha_1, \ldots, \alpha_n \rangle \in \{0,1\}^{n \times n}.$$

For all $i \in \{1, \ldots, n\}$,

$$\text{if} \begin{cases} a_i = 0 & \text{then} \quad V \text{ checks that } f(\alpha_i) = x_i' \\ a_i = 1 & \text{then} \quad V \text{ checks that } f(\alpha_i) = y_i' \end{cases}$$

If any of these n equalities fail then V rejects, else V accepts. We refer to this process as "V uses α to verify that c is committed to a".

———∞———

We now describe the three phases of a many messages signature scheme. In the initialization phase, S generates an index of a universal one-way hash function and a pointer component. The universal one-way hash function is used to compress the public part of each block from $4n^2$ bits down to n bits, and the pointer component is used to commit to the compression of the public part of the first block.

Initialization : S randomly chooses an index $z \in_{\mathcal{U}} \{0,1\}^{\ell(n)}$ for the universal one-way hash function. S creates a the pointer component $p_0 = \langle ps_0, pv_0 \rangle$. S send $\langle z, pv_0 \rangle$ to V using a public line.

Note that at this point in time, since S uses a public line to send to V the values z and pv_0, V is sure that $\langle z, pv_0 \rangle$ was generated by S.

Signing a message : When the signer S wants to sign the ith message $m_i \in \{0,1\}^n$, S creates a block and links it into the list as described below. Suppose that S has already signed messages

$$m_1, \ldots, m_{i-1} \in \{0,1\}^n.$$

Then, S has already created, for all $j \in \{1, \ldots, i-1\}$, block $b_j = \langle p_j, d_j \rangle$, where p_j is the pointer component and d_j is the data component. For any j, let $p_j = \langle ps_j, pv_j \rangle$ be the pointer component of the jth block, let $d_j = \langle ds_j, dv_j \rangle$ and let $v_j = \langle pv_j, dv_j \rangle$ be the concatenation of the public parts of the jth components. For all $j \in \{1, \ldots, i-1\}$, S has already committed d_j to sign message m_j. For all $j \in \{0, \ldots, i-2\}$, S has already used p_j to commit to $g_z(v_{j+1})$. However, S has not used p_{i-1}.

Here is how S signs m_i. S creates components p_i and d_i. S uses d_i to commit to m_i. Let α_i be the commitment of d_i to m_i. S uses p_{i-1} to commit to $g_z(v_i)$. Let β_i be the commitment of p_{i-1} to $g_z(v_i)$. The information sent by S to V on a public network is

$$\langle m_i, v_i, \alpha_i, \beta_i \rangle.$$

Verifying a message : At the point when V is to verify the signature of m_i, V has all the public information sent by S. The part of this information used to verify the signature consists of:

- The index z of the universal one-way hash function g used to compress v_j for all $j \in \{1, \ldots, i\}$.

- For all $j \in \{1, \ldots, i\}$, the public parts $v_j = \langle pv_j, dv_j \rangle$ of the pointer and data components of block j.

- For all $j \in \{0, \ldots, i-1\}$, the commitment β_j of p_j to $g_z(v_{j+1})$.

- The commitment α_i of d_i to m_i.

Here is how V verifies that S signed m_i. For all $j \in \{0, \ldots, i-1\}$, V checks to see if p_j is committed to $g_z(v_{j+1})$. This consists of computing the value of $g_z(v_{j+1})$ and then checking that β_j is a commitment of p_j to $g_z(v_{j+1})$. V checks to see if α_i is a commitment of d_i to m_i. If all checks are equalities then V accepts, else V rejects.

———————∞———————

The security parameter $s(n)$ of the scheme is $4n^2 \cdot (i + 1)$, where i is the total number of messages sent. This completes the description of the signature scheme. We now show that this signature scheme is secure against adaptive chosen message attack.

Theorem 17.1 : If f is a one-way permutation then the scheme is secure against adaptive chosen message attack. The reduction is weak-preserving.

PROOF: (Sketch) Let A be an adversary that runs in time $T(n)$ and has success probability $\delta(n)$ when it mounts an adaptive chosen message attack against the above signature scheme. We informally describe an oracle adversary M such that $M^{A,S,V}$ can invert $y = f(x)$ for $x \in_{\mathcal{U}} \{0,1\}^n$. During the course of the attack run by A, the signer S privately generates a polynomial number of inputs to the one-way permutation f, reveals to A the value of f applied to all of these inputs, but only reveals to A a subset of the inputs interactively to A as the attack progresses, keeping some inputs private. The way $M^{A,S,V}$ works is to generate on its own all but one of these inputs that S generates and apply f to them, and use y as the value of f applied to the remaining input. The key point is that in the first part of the attack when A is interacting with S, there is a good chance that the inverse of y with respect to f will not be required of S. On the other hand, in the second part of the attack when A is interacting with V, if A is able to convince V to accept a message not signed by S in the first part of the attack, then A is able to generate one of the inverses not revealed in the first part of the attack, and with some probability this is the inverse of y.

We now describe the attack A mounts. In the initialization phase, both V and A receive pv_0 and z from S. Then, A uses S to sign messages $m_1, \ldots, m_i$, and from this A receives the following information:

- For all $j \in \{1, \ldots, i\}$, the public parts $v_j = \langle pv_j, dv_j \rangle$ of the pointer and data components of block j.

- For all $j \in \{0, \ldots, i - 1\}$, the commitment β_j of p_j to $g_z(v_{j+1})$.

- The commitment α_i of d_i to m_i.

Finally, based on the information received from S, A generates a message $m' \notin \{m_1, \ldots, m_i\}$ and information that is supposed to convince that m' was signed by S. Without loss of generality, we assume that A attempts to convince V that m' is the ith message signed by S. The information A sends to V includes m', the purported public parts $v'_1, \ldots, v'_i$, the purported commitment α' to m', and the purported commitments $\beta'_0, \ldots, \beta'_{i-1}$.

Suppose that V accepts that m' was signed by S based on $\langle z, pv_0 \rangle$ and on the information that A sends to V. For all $j \in \{0, \ldots, i-1\}$, let $a_j = g_z(v_{j+1})$ and let $a'_j = g_z(v'_{j+1})$. Let

$$k = \min\{j \in \{0, \ldots, i\} : \langle v'_j, a'_j \rangle \neq \langle v_j, a_j \rangle\}.$$

- If k is undefined, i.e., there is no such j, then in particular $dv'_i = dv_i$. Since α' is a valid commitment of dv_i to m' and S only revealed to A a valid commitment of dv_i to $m_i \neq m'$, α' contains an inverse of one of the random outputs of f generated by S which was not revealed to A. With some significant probability, this is the inverse of y, and thus f is successfully inverted.

- If k is defined and $a'_k \neq a_k$ then in particular $pv'_{k-1} = pv_{k-1}$. Since β'_{k-1} is a valid commitment of pv_{k-1} to a'_k and S only revealed to A a valid commitment of pv_{k-1} to $a_k \neq a'_k$, β'_{k-1} contains an inverse of one of the random outputs of f generated by S which was not revealed to A. With some significant probability, this is the inverse of y, and thus f is successfully inverted.

- If k is defined and $a'_k = a_k$ and $v'_k \neq v_k$ then the adversary A was able to find v'_k such that $g_z(v'_k) = h_z(v_k)$ where $v'_k \neq v_k$ and where v_k was generated by S independently of z. With some significant probability, y can be inverted as described in the proof of Theorem 16.2.

■

Exercise 67 : Formally prove Theorem 17.1.   ♠

Solutions to the next couple of exercises show that the scheme described above is not as efficient as it could be in more than one way.

One problem with the multiple signature scheme just described is that it is in the form of a linked list, where each element in the linked list is a window. A solution to the following exercise shows that there is a more efficient way to build the signature scheme data structure.

Exercise 68 : Show how to construct a multiple messages signature scheme that uses a balanced tree structure, i.e., the length of a path in the data structure to sign the ith is of length at most $\log(i)$.   ♠

Another problem with the multiple signature scheme just described is that the amount of space that the signer has to remember is not simply a function of the security parameter, but also of the number of messages signed. A solution to the following exercise shows that this is not an inherent problem.

Exercise 69 : Show how to build a multiple messages signature scheme such that the memory needed by the signer is $n^{\mathcal{O}(1)}$, and doesn't depend on the number of messages signed.

Hint : Consider the tree construction idea used to construct a pseudo-random function generator from a pseudorandom generator as described in Lecture 12 on page 123. ♠

Lecture 18

Overview

We define interactive proof systems. We give examples of languages which have **IP** proofs but which are not known to be in **NP**. We define and give a construction for a hidden bit commitment scheme. We define zero knowledge interactive proofs and describe a computational secure zero knowledge interactive proof for all languages in **NP** based on a hidden bit commitment scheme.

NP viewed as restricted IP

IP, which stands for interactive proof, is a complexity class that is a generalization of **NP**. To compare the two complexity classes, we first briefly review some properties of **NP**. If a language $L \in$ **NP** then there is a **P**-time function ensemble $M : \{0,1\}^n \times \{0,1\}^{\ell(n)} \to \{0,1\}$ such that, for all $x \in \{0,1\}^n$, $x \in L$ iff there is some $w \in \{0,1\}^{\ell(n)}$ such that $M(x,w) = 1$. Three fundamental aspects of **NP** are the following. For all $x \in \{0,1\}^n$,

(Completeness): If $x \in L$, then there is a witness $w \in \{0,1\}^{\ell(n)}$ such that $M(x,w) = 1$.

(Soundness): If $x \notin L$, then there is no witness $w \in \{0,1\}^{\ell(n)}$ such that $M(x,w) = 1$.

(Efficiency): If $w \in \{0,1\}^{\ell(n)}$ is a witness to $x \in L$, then membership in L can be certified in $n^{\mathcal{O}(1)}$ time given w by computing $M(x,w)$.

Now we give an alternative way of viewing **NP**.

Definition (restricted IP): Let $\langle P, V \rangle$ be a pair of **TM**s, called the prover and verifier, respectively. Both P and V have as input $x \in \{0,1\}^n$. P does some computation and sends some information to V. Based on this, V does some computation and either accepts (outputs 1) or rejects (outputs 0). Let out$(\langle P, V \rangle, x)$ denote the output of the verifier V with respect to prover P on input x. A language L is in *restricted* **IP** if the protocol satisfies the following conditions:

(Completeness): If $x \in L$, then out$(\langle P, V \rangle, x) = 1$, i.e., P convinces V that $x \in L$.

(Soundness): If $x \notin L$, then for all provers P', out($\langle P', V \rangle, x) = 0$, i.e., there is no way to convince V that $x \in L$.

(Efficiency): V is a **P**-time function ensemble. We allow provers unlimited resources. ♣

Exercise 70 : Prove that **NP** = restricted **IP**. ♠

Definition (IP): **IP** is the generalization of restricted **IP** where we allow two additional resources:

(Randomness): V is a randomized **P**-time function ensemble.

(Interaction): P and V communicate back and forth several times. There are $p(n) = n^{\mathcal{O}(1)}$ rounds of interaction. In each round P sends some information to V and V sends some information to P.

Because the prover has unbounded time resources, allowing the prover to be randomized doesn't add computational power. However, many of the provers we describe run in $n^{\mathcal{O}(1)}$ time when given as input an advice string of length $n^{\mathcal{O}(1)}$. The definition of efficiency remains the same, but our notions of completeness and soundness are changed to be probabilistic, similar to the definition of **BPP**. There are constants $0 \leq c_0 < c_1 \leq 1$ such that:

(Completeness): If $x \in L$ then $\Pr[\text{out}(\langle P, V \rangle, x) = 1] \geq c_1$.

(Soundness): If $x \notin L$ then, for all P', $\Pr[\text{out}(\langle P', V \rangle, x) = 1] \leq c_0$. ♣

Exercise 71 : Prove that you can increase the probability spread between the completeness and soundness conditions by repeating the protocol several times. ♠

Relationship of IP to familiar classes

It is clear that **NP** $\subseteq$ **IP**. Also, **BPP** $\subseteq$ **IP**, because if $L \in$ **BPP** then we can construct an **IP** protocol for L by simply letting the verifier V be the randomized **P**-time function ensemble associated with L as described in the definition of **BPP**, and V ignores whatever the prover sends on the conversation tapes and decides membership in L on its own. A series of results show that **IP** = **PSPACE**, where **PSPACE** is the complexity class consisting of all polynomial space computations.

Definition (graph isomorphism):   Let $G = (N, E)$ be a graph on node set $N = \{1, \dots, n\}$ and edge set E. For all $\pi \in \textbf{Perm}{:}N \to N$, let $\pi(G)$ be the graph obtained by relabeling each node $i \in N$ with $\pi(i)$. Two graphs $G_0 = (N, E_0)$ and $G_1 = (N, E_1)$ are said to be *isomorphic* if there is a permutation $\pi \in \textbf{Perm}{:}N \to N$ such that $\pi(G_0) = G_1$. We let $G_0 \cong G_1$ denote that G_0 and G_1 are isomorphic graphs.   ♣

Here is an example of a language that is easily seen to be in **IP**, but whose relationship to **NP** is unknown.

Graph non-isomorphism language GNI :   Let $N = \{1, \dots, n\}$ and let $\langle G_0, G_1 \rangle$ be a pair of undirected graphs $G_0 = (N, E_0)$ and $G_1 = (N, E_1)$. Then, $\langle G_0, G_1 \rangle \in$ GNI if and only if $G_0 \ncong G_1$.

Theorem 18.1 :   GNI $\in$ **IP**.

PROOF: We describe the **IP** protocol on input $\langle G_0, G_1 \rangle \in$ GNI.

(1) V chooses $b \in_{\mathcal{U}} \{0,1\}$ and $\pi \in_{\mathcal{U}} \textbf{Perm}{:}N \to N$. V computes $H = \pi(G_b)$, sends H to P.

(2) P computes $b' \in \{0,1\}$ such that there is a $\pi' \in \textbf{Perm}{:}N \to N$ with $\pi'(G_{b'}) = H$, sends b' to V.

(3) V accepts if $b = b'$ and rejects if $b \neq b'$.

(**Completeness**): Suppose $G_0 \ncong G_1$. Since $H \cong G_b$ it follows that $H \ncong G_{1-b}$, and thus P produces b' with $b' = b$ and V accepts.

(**Soundness**): Suppose that $G_0 \cong G_1$. Then, H and b are independently distributed. Since b' only depends on H, $b = b'$ with probability $1/2$ independent of what a prover P' does in step (2), and thus V rejects with probability $1/2$.

(**Efficiency**): V is a randomized **P**-time function ensemble.

■

Theorem 18.1 is interesting because it is still open whether GNI $\in$ **NP**.

Zero Knowledge

The motivation behind a zero knowledge interactive proof (ZKIP) is natural: Suppose you (the prover) have a marvelous proof of Fermat's last theorem, [1] and you would like to convince a colleague (the verifier) that you have a proof, but you don't want to reveal any information about *how* you proved it in the process, since you don't really trust this colleague. In such a case you would want to execute a protocol that convinces the colleague you know the proof, but reveals nothing to the colleague except that you know the proof.

How do we formalize the requirement that the protocol reveals no information? Let L be a language that we would like to show is in **ZKIP**. The protocol has the property that for all randomized **P**-time function ensemble V' there is a randomized **P**-time function ensemble S' such that on input $x \in L$, S' on its own produces all information seen by V' interacting with P in the actual protocol. Thus, whatever information V' receives from the protocol can be produced without any help from P; and thus the only information V' gains from the interaction with P is confidence that $x \in L$.

We allow the verifier V to have an auxiliary input $y \in \{0,1\}^{p(n)}$ in addition to the common input x, where y can be thought of as all the information that V had before the start of the protocol (perhaps y is the information gained from previous interactions with P or another prover P' in some other protocol). The reason for the auxiliary input is that zero knowledge interactive proofs are often used within cryptographic protocols between two parties P and V in the following way. During an intermediate stage of the protocol, V requires P to convince V that P is following the protocol before V is willing to continue with the rest of the protocol. However, P doesn't want to leak any extra information beyond the fact that P is following the protocol to V. The protocol is designed in such a way that P and V have generated together a common string x, and P is following the protocol if $x \in L$. The auxiliary string y in this case is information that V has computed from the previous parts of the protocol.

Definition (view of V'): Let r be the bits produced by the source of randomness used by V', let $x \in \{0,1\}^n$ be the common input to P and V' and let $y \in \{0,1\}^{p(n)}$ be the auxiliary input to V'. Define view($\langle P, V' \rangle, x, y$) as $\langle x, y, r \rangle$ concatenated with all the information sent back and forth between P and V' when running the **IP** protocol.   ♣

The view of V' completely determines the behavior of V' during the protocol. Note that view($\langle P, V' \rangle, x, y$) defines a distribution.

Definition (ZKIP): We say $L \in$ **ZKIP** if $L \in$ **IP** and the proto-

---

[1] Recently, A. Wiles proved Fermat's last theorem, and it turns out that he revealed the proof to a few of his colleagues several months before announcing it to the world.

col showing $L \in$ **IP** is zero knowledge. Three different notions of zero knowledge are the following. For every V', for every $x \in L \cap \{0,1\}^n$, for every $y \in \{0,1\}^{p(n)}$, the two distributions $\text{view}(\langle P, V' \rangle, x, y)$ and $S'(x, y)$ are

(perfect ZKIP): identical.

(statistical ZKIP): at most $\epsilon(n)$- statistically distinguishable.

(computational ZKIP): $\mathbf{S}(n)$-computationally indistinguishable. ♣

We only consider perfect and computational zero knowledge in the remainder of this lecture.

An important point is that there is no zero knowledge requirement when $x \notin L$. The intuition for this is that P is trying to convince V that $x \in L$, but if it turns out that $x \notin L$ then P shouldn't be trying to convince V of this in the first place, and hence there is no need for protection against the protocol leaking information.

Question :   Is the **IP** protocol for GNI a zero knowledge protocol?

Answer :   If the verifier is V then when $G_0 \not\cong G_1$ the proof is zero knowledge (this is easy to see – the simulator just sends back $b' = b$). However, we must make sure it is zero knowledge for *all* potential verifiers V'. The protocol described before is not zero knowledge if graph isomorphism is not in **P**. For example, suppose that a verifier V' has graph G_2 which is known to be isomorphic to one of G_0 or G_1. Then by sending $H = G_2$ to P, V' can find out which one it is (something V' or a simulator can't do alone efficiently without an efficient algorithm for graph isomorphism).

———————∞———————

A simple example of a language with a zero knowledge interactive proof is the graph isomorphism language.

Graph isomorphism language GI :   Let $N = \{1, \ldots, n\}$ and let $\langle G_0, G_1 \rangle$ be a pair of undirected graphs $G_0 = (N, E_0)$ and $G_1 = (N, E_1)$. Then, $\langle G_0, G_1 \rangle \in$ GI if and only if $G_0 \cong G_1$.

Theorem 18.2 :   GI has a perfect **ZKIP**.

PROOF: The common input to the **IP** protocol is $x = \langle G_0, G_1 \rangle \in$ GI and the auxiliary input y for V.

(1) P chooses $\pi \in_{\mathcal{U}}$ **Perm**:$N \to N$, computes $H = \pi(G_0)$, sends H to V.

(2) V chooses $b \in_{\mathcal{U}} \{0, 1\}$, sends b to P.

(3) P finds the permutation π' such that $\pi'(G_b) = H$, sends π' to V.

(4) V accepts if $\pi'(G_b) = H$ and otherwise V rejects.

The protocol is complete because if $G_0 \cong G_1$ then H is isomorphic to both G_0 and G_1, so for both values of b, P can find a permutation π' such that $\pi'(G_b) = H$. The protocol is sound since if $G_0 \ncong G_1$, then, independent of what a prover P' does in step (1), with probability $1/2$ it will happen that $G_b \ncong H$, and thus with probability $1/2$, P' will not be able to produce π' such that $\pi'(G_b) = H$ and thus V will not accept. It is easy to verify the protocol is efficient.

We now show the protocol is zero knowledge. Consider simulating a run of the protocol between prover P and randomized **P**-time function ensemble V' with common input $x = \langle G_0, G_1 \rangle$, where $G_0 \cong G_1$, and auxiliary input y for V'. The simulating randomized **P**-time function ensemble S' first randomly fixes the bits r used by V' as its source of randomness. S' simulates step (1) by choosing $b' \in_{\mathcal{U}} \{0, 1\}$ and $\pi \in_{\mathcal{U}}$ **Perm**:$N \to N$, setting $H = \pi(G_{b'})$, and sending H to V'. Then S' simulates V' in step (2), and V' produces b.[2] If $b = b'$ then S' in step (3) sets $\pi' = \pi$ and sends π' to V'. In this case, we say that the simulation is successful. If $b \neq b'$ then S' cannot simulate step (3) correctly without computing an isomorphism between G_0 and G_1. In this case we say that the simulation is unsuccessful. S' in this case backs up the computation of V' to its state just at the beginning of step (1) (S' can do this because S' has complete control over V'), and continues the simulation from there, but uses new random bits to make the choices for b' and π. S' continues this backing up process until there is a successful simulation. The output of S' is $\langle x, y, r, H, b, \pi \rangle$, where $\langle H, b, \pi \rangle$ is the information sent in the successful simulation.

We now show that S' runs in expected $n^{\mathcal{O}(1)}$ time. Since $G_0 \cong G_1$ and $\pi \in_{\mathcal{U}}$ **Perm**:$N \to N$, H is a random permutation of both G_0 and G_1, and b' is independent of H. Thus, since V' only sees H from step (1)

---

[2]Note that V' may be doing something completely different than what V does in two possible ways, one easily detectable and the other not. It is easily detectable by P if V' deviates in format from the protocol, e.g., if V' doesn't produce $b \in \{0, 1\}$ in step (2). If V' deviates in this way from the protocol then P immediately halts the protocol. Without loss of generality, we assume V' doesn't deviate in format from the protocol. The other kind of deviation by V' is impossible for P to catch; V' may be doing arbitrary computations and may for example not choose b randomly, although $b \in \{0, 1\}$. It is more difficult to guarantee that V' doesn't gain information from these kinds of deviations, and this is the substance of the zero knowledge part of the proof.

of the simulation (and not the value of b'), $b = b'$ with probability $1/2$. (This is the same argument as was used to prove soundness for Theorem 18.1.) Since $b = b'$ with probability $1/2$ the expected number of times S' has to back up before a successful simulation is two.

We now show that $S'(x,y) = \text{view}(\langle P, V' \rangle, x, y)$. Once r is fixed, the behavior of V' only depends on the information received from P as simulated by S'. For each possible graph H that S' sends to V' in step (1) it is equally likely that H came from G_0 or G_1, i.e., that $b' = 0$ or $b' = 1$. Given H, V' deterministically outputs either $b = 0$ or $b = 1$ in step (2). For exactly one of the two ways that S' could have chosen H the simulation is successful ($b = b'$) and for the other the computation has to be backed up ($b \neq b'$). It follows that the distribution on H among successful simulations is $\Pi(G_0)$, where $\Pi \in_{\mathcal{U}}$ **Perm**$:N \rightarrow N$. This is exactly the same distribution on H as in the protocol between V' and P. Since the computation of π' by P in step (2) is completely determined by H and b, and it is exactly the same π' produced by S' on successful simulations, the claim follows.   ∎

Consider the protocol where step (1) is replaced with "P sends to V the graph $H = G_0$". It can be shown that this protocol is an **IP** protocol for GI, but it certainly is not a **ZKIP** protocol. Thus, the crucial step, where the prover P protects the zero knowledge aspect of the proof, is at the random choice of a permutation in step (1).

The only place where P is required to do more than a polynomial amount of computation is in step (3). If P knows the isomorphism between G_0 and G_1, i.e., π'' such that $\pi''(G_0) = G_1$, then step (3) can be performed by P in polynomial time.

Exercise 72 : Show that GNI has a perfect **ZKIP**.   ♠

Hidden Bit Commitment

We now introduce hidden bit commitment schemes and show that a hidden bit commitment scheme can be constructed based on a one-way function. In the next section, we briefly sketch the proof that any **NP** language has a computational zero knowledge proof based on a hidden bit commitment scheme.

A hidden bit commitment consists of two phases. In the commit phase a party S sends information to a party V that commits S to a bit $b \in_{\mathcal{U}} \{0, 1\}$ without revealing to V any information about the value of b. In the release phase S sends information to V that convinces V that S truly committed to a particular value of b in the commit phase.

Definition (hidden bit commitment): Let S and V be randomized **P**-time function ensembles. Let n be the security parameter of the protocol, i.e., n is the total amount of information that S keeps hidden from V in the commit phase.

(commit phase) S chooses $b \in_{\mathcal{U}} \{0,1\}$ and then interacts with V.

(release phase) S sends to V the bit b and then interacts with V. At the end, V either accepts or rejects. ♣

Definition (hidden bit): Let A be an adversary that takes the place of V in the commit phase of the protocol. The success probability of A is the correlation between b and the output of A. The protocol is $\mathbf{S}(n)$-*secure hidden* if every adversary has time-success ratio at least $\mathbf{S}(n)$. ♣

Definition (committed bit): Let A be an adversary that takes the place of S as follows. First, A interacts with V in the commit phase. Then A interacts with V in the release phase twice (both times starting with V at the same point) the first time to try and convince V that the committed bit is 0 and the second time to try and convince V that the committed bit is 1. The success probability of A is the probability that A is able to convince V both times. The protocol is $\mathbf{S}(n)$-*secure committed* if every adversary has time-success ratio at least $\mathbf{S}(n)$. ♣

Note that a hidden and committed bit is exactly what a one-way predicate provides. (See page 149.)

Definition (secure hidden bit commitment): The hidden bit commitment protocol is $\mathbf{S}(n)$-secure if it is $\mathbf{S}(n)$-secure hidden and $\mathbf{S}(n)$-secure committed. ♣

The following protocol shows how S and V may use a hidden bit commitment protocol to toss a coin using a public line so that no party can bias the outcome.

Coin tossing using a public line :

- S chooses $b \in_{\mathcal{U}} \{0,1\}$. S and V run the commit phase to commit S to b.

- V sends $b' \in_{\mathcal{U}} \{0,1\}$ to S.

- S sends b to V.

- S and V run the release phase to show that S is committed to b.

The bit agreed upon is $b \oplus b'$.

————∞————

The main property of the coin tossing protocol is that, as long as both parties are guaranteed to complete the protocol, the bit that they agree on is unbiased, in the following sense. The guarantee is that if party P enacts its portion of the protocol as specified then then agreed upon bit is a random bit. This provides P with protection against possible cheating by the other party. Of course, if both parties try to cheat simultaneously, then the agreed upon bit may not be at all random.

One can directly construct a hidden bit commitment protocol using a one-way predicate (page 149). On page 149 we show how to construct a one-way predicate from any one-way permutation. We now describe an alternative method that is based on a pseudorandom generator.

A hidden bit commitment protocol :

Let $g : \{0,1\}^n \rightarrow \{0,1\}^{3n}$ be a pseudorandom generator with security parameter n.

Commit phase :

- V chooses $r \in_{\mathcal{U}} \{0,1\}^{3n}$, sends r to S.

- S picks $b \in_{\mathcal{U}} \{0,1\}$ and $x \in_{\mathcal{U}} \{0,1\}^n$. S computes the string $c \in \{0,1\}^{3n}$, where, for all $i \in \{1,\ldots,3n\}$,

$$c_i = \begin{cases} g(x)_i & \text{if } r_i = 0 \\ g(x)_i \oplus b & \text{if } r_i = 1 \end{cases} .$$

  S sends c to V.

Release phase :

- S sends x and b to V.

- V verifies, for all $i \in \{1,\ldots,3n\}$, that

$$g(x)_i = \begin{cases} c_i & \text{if } r_i = 0 \\ c_i \oplus b & \text{if } r_i = 1 \end{cases} .$$

  V accepts if all equalities are verified.

————∞————

The bit b is statistically committed but only hidden to computationally limited adversaries. There are alternative protocols for hidden bit commitment, not discussed in this monograph, where the bit is committed to computationally limited adversaries but statistically hidden.

Exercise 73 : Prove that if g is a pseudorandom generator then the hidden bit commitment protocol hides b. The reduction should be linear-preserving. Prove that the protocol is 2^n-secure committed, independent of whether or not g is a pseudorandom generator. ♠

Computational ZKIP for NP

Hamilton cycle language HC : Let $N = \{1, \ldots, n\}$ and let $G = (N, E)$ be an undirected graph, Then, $G \in \text{HC}$ if and only if there is Hamilton cycle in G, i.e., a cycle in G that visits each vertex exactly once. Let $M(G)$ be the $n \times n$ adjacency matrix for G, i.e., $M(G)_{i,j} = 1$ if $(i, j) \in E$ and $M(G)_{i,j} = 0$ otherwise. Then, there is a Hamilton cycle in G if and only if there is a cyclic permutation $\pi \in \textbf{Perm}:N \to N$ such that, for all $i \in N$, $M(G)_{i,\pi(i)} = 1$.

————∞————

HC is **NP**-complete, and furthermore the standard reduction of a language $L \in \textbf{NP}$ to HC preserves witnesses, i.e., if $x \in L$ and w is a witness for x then given $\langle x, w \rangle$ the reduction produces a graph G together with a Hamilton cycle h in G. Thus, to show that every **NP** language has a computational **ZKIP**, it is enough to show that HC has a computational **ZKIP**.

Theorem 18.3 : HC has a computational **ZKIP**.

PROOF: (Sketch) We describe the **IP** protocol on input $G \in \text{HC}$.

(1) P chooses $\pi \in_{\mathcal{U}} \textbf{Perm}:N \to N$ and computes $M' = M(\pi(G))$. P executes the commit phase of the hidden bit commitment scheme and sends V the commitment to all the bits in M'.

(2) V randomly chooses $b \in_{\mathcal{U}} \{0, 1\}$, where $b = 0$ is interpreted as "Show me the permutation π," and $b = 1$ is interpreted as "Show me a Hamilton cycle for M'." V sends b to P.

(3) If $b = 0$ then P sends π to V and uses the release phase of the hidden bit commitment scheme to release all the bits of M'. If $b = 1$ then P sends the Hamilton cycle $h' = \pi(h)$ to V and uses the release phase of the hidden bit commitment scheme to release all the bits corresponding to h' in M'.

(4) If $b = 0$ then V checks that $M(\pi(G)) = M'$ and accepts if true. If $b = 1$ then V checks that all the bits in h' are set to 1 in M' and that h' is a cyclic permutation and accepts if true. In all other cases, V rejects.

When P has a Hamilton cycle h in G, by following the protocol P can always make V accept, and thus the protocol is complete. Conversely, it is easy to see that when there is no Hamilton cycle in G, then every prover P' will be unable to satisfactorily answer one of the two queries, $b = 0$ or $b = 1$. Since P commits to M' before learning which query will be asked, with probability $1/2$ P will not be able to answer the chosen query, in which case V rejects. Thus, the protocol is sound. The protocol is easily seen to be efficient.

Now we show the computational zero knowledge property. Given a randomized **P**-time function ensemble V', we construct a simulator S' similar to the simulator in the proof of Theorem 18.2 (page 178). S' in step (1) chooses $b' \in_{\mathcal{U}} \{0, 1\}$ and then commits to an M' of type b' as described below.

(type 0) S' chooses $\pi \in_{\mathcal{U}}$ **Perm**:$N \to N$ and commits to $M' = M(\pi(G))$.

(type 1) S' chooses a random cyclic permutation $\pi' \in$ **Perm**:$N \to N$ and commits to the matrix M' with all entries equal to 0 except that $M'_{i,\pi'(i)} = 1$ for all $i \in N$. (This corresponds to the graph with edges that correspond to a random Hamilton cycle.)

If $b = b'$ then S' can continue the simulation in step (3) and the simulation is considered successful. If $b \neq b'$ then S' backs up the simulation to step (1) and tries again (the same idea as in the proof of Theorem 18.2). The probability that $b = b'$ is about $1/2$, because if it is much different than $1/2$ then V' can tell the difference between a commitment to a 0 bit and a commitment to a 1 bit, and this gives a way of inverting the one-way permutation on which the security of the hidden bit commitment scheme is based. The rest of the proof is similar to the proof of Theorem 18.2. ∎

In the protocol, if the prover P knows a Hamilton cycle h in G then P can be a randomized **P**-time function ensemble.

List of Exercises and Research Problems

Note : Exercises marked **(easy)** are meant to test immediate understanding of the material. The listing below is mainly to be used as a quick reference guide. In some cases, the complete description of the exercise is not given below, and/or it is taken out of context. See the original exercise for a complete description.

Preliminaries

Prove that the answer to the decision version of the $\mathbf{P} = \mathbf{NP}$ question is yes if and only if the answer to the search version of the question is yes.

Prove that $\mathbf{BPP}(1/n^c) \subseteq \mathbf{BPP}(\epsilon) \subseteq \mathbf{BPP}(1 - 1/2^{n^c})$,

Prove that $\mathbf{RP} \subseteq \mathbf{P/poly}$ and $\mathbf{BPP} \subseteq \mathbf{P/poly}$.

Given X (not necessarily ≥ 0) such that $E[X] = \mu$ and $X \leq 2\mu$, give an upper bound on $\Pr[X < \frac{\mu}{2}]$

Let $X, X_1, \ldots, X_n$ be identically distributed and pairwise independent $\{0, 1\}$-valued random variables and let $p = \Pr[X = 1]$. Prove using Chebychev's inequality that:

$$\Pr\left[\left|\frac{1}{n}\sum_{i=1}^{n} X_i - p\right| \geq \delta\right] \leq \frac{p(1-p)}{\delta^2 n}.$$

Lecture 1

Show that $\mathbf{P} = \mathbf{NP}$ implies there are no one-way functions.

Let $A \in_\mathcal{U} \{0, 1\}^n$ and let $B \in_\mathcal{U} \{0, 1\}^{n \times (n+1)}$. Prove the probability that $f(A, B) = \langle \sum_{i=1}^{n} A_i \cdot B_i, B \rangle$ has a unique inverse is lower bounded by a constant strictly greater than zero independent of n.

Show that $\mathbf{P} = \mathbf{NP}$ implies there are no pseudorandom generators.

Lecture 2

Exercise 9 26
Let $f : \{0,1\}^n \to \{0,1\}^{\ell(n)}$ be a one-way function and let $X \in_\mathcal{U} \{0,1\}^n$. Show that for any adversary A there is an adversary A' with worst case time-success ratio at most n times the average case time-success ratio of A.

Lecture 3

Research Problem 1 48
Design a linear-preserving (or poly-preserving) reduction from an arbitrary weak one-way function to a one-way function.

Research Problem 2 48
Design a linear-preserving (or poly-preserving) reduction from weak one-way permutation f to one-way permutation g such that the parallel time for computing g is comparable to that for computing f.

Lecture 4

Exercise 10 51
Prove the first part of Theorem 4.1.

Exercise 11 51
Prove the second part of Theorem 4.1.

Exercise 12 54
Prove that if there is a $\mathbf{S}(n)$-secure pseudorandom generator then $\mathbf{BPP} \subseteq \mathbf{DTIME}(\mathcal{S}(n))$.

Research Problem 3 55
Design a linear-preserving (or poly-preserving) reduction from a one-stretching pseudorandom generator to an n-stretching pseudorandom generator that can be computed fast in parallel.

Lecture 5

Exercise 13 (easy) 57
Generalize the Modulo Prime Space to a probability space where

$$X_0, \ldots, X_{m-1} \in_\mathcal{U} \mathcal{Z}_p$$

are k-wise independent, where the size of the probability space is p^k.

Exercise 14 (easy) 59
Prove the pairwise independence property for the Inner Product Space.

Describe a natural hybrid between the two methods that uses $2k\ell(n)$ random bits and

$$m = \max\left\{ \lceil 2/\epsilon \rceil \cdot \left\lceil (1/\delta)^{1/k} \right\rceil, \lceil k/\epsilon \rceil \right\}$$

witness tests.

Find an $\mathcal{O}(m + n)$ time algorithm for the Vertex Partition Problem.

Find a parallel algorithm that uses $\mathcal{O}(m+n)$ processors and runs in time $\mathcal{O}(\log^2(m + n))$ for the Vertex Partition Problem.

Let p be a positive integer and let $X_1, \ldots, X_n \in_{\mathcal{U}} \mathcal{Z}_p$ be a sequence of four-wise independent random variables. Define random variable

$$Y = \min\{|X_i - X_j| : 1 \le i < j \le n\}.$$

Prove that there is a constant $c > 0$ such that for any $\alpha \le 1$

$$\Pr[Y \le \alpha p/n^2] \ge c\alpha.$$

Lecture 6

From the Hidden Bit Theorem, show that if $f(x)$ is a one-way permutation then $g(x, z) = \langle f(x), x \odot z \rangle$ is a pseudorandom generator. The reduction should be poly-preserving.

Let $X \in \{0, 1\}^n$. Describe a one-way permutation $f : \{0, 1\}^n \to \{0, 1\}^{\ell(n)}$ where X_1 is not hidden given $f(X)$. Let $f : \{0, 1\}^n \to \{0, 1\}^{\ell(n)}$ be a **P**-time function ensemble and let $I \in_{\mathcal{U}} \{1, \ldots, n\}$. Show that if X_I can be predicted with probability greater than $1 - 1/(2n)$ given $f(X)$ then f is not a one-way function.

Describe a **P**-time function ensemble $f : \{0, 1\}^n \to \{0, 1\}^{\ell(n)}$ which is certainly not a one-way function but for which the inner product bit is provably 2^n-secure.

Lecture 7

Exercise 22 (easy) 70
Let $f : \{0,1\}^n \to \{0,1\}^{\ell(n)}$ be a function ensemble. Show that

$$\text{dist}(f(X), f(Y)) \leq \text{dist}(X, Y).$$

Exercise 23 (easy) 70
Describe a statistical test t such that $\delta_t(n) = \text{dist}(\mathcal{D}_n, \mathcal{E}_n)$.

Exercise 24 (easy) 71
Prove that for any triple of distributions $\mathcal{D}_n^1 : \{0,1\}^n$, $\mathcal{D}_n^2 : \{0,1\}^n$, and $\mathcal{D}_n^3 : \{0,1\}^n$, $\text{dist}(\mathcal{D}_n^1, \mathcal{D}_n^3) \leq \text{dist}(\mathcal{D}_n^1, \mathcal{D}_n^2) + \text{dist}(\mathcal{D}_n^2, \mathcal{D}_n^3)$.

Exercise 25 71
Let $f : \{0,1\}^n \to \{0,1\}^{\ell(n)}$ be a function ensemble that can be computed in time $n^{\mathcal{O}(1)}$ on average, i.e., for $X \in_{\mathcal{U}} \{0,1\}^n$, $\mathbf{E}_X[T(X)] = n^{\mathcal{O}(1)}$. where $T(x)$ is the time to compute f on input x. Show that for any $m(n) = n^{\mathcal{O}(1)}$ there is a $p(n) = n^{\mathcal{O}(1)}$ and a **P**-time function ensemble $f' : \{0,1\}^{p(n)} \to \{0,1\}^{\ell(n}$ such that $\text{dist}(f(X), f'(Z)) \leq 1/m(n)$, where $Z \in_{\mathcal{U}} \{0,1\}^{p(n)}$.

Exercise 26 (easy) 71
Prove that if $\mathcal{D}_n$ and $\mathcal{E}_n$ are two probability ensembles that are at most $\epsilon(n)$-statistically distinguishable then $\mathcal{D}_n$ and $\mathcal{E}_n$ are $(1/\epsilon(n))$-secure computationally indistinguishable.

Exercise 27 71
Prove that if $\mathcal{D}_n^1$ and $\mathcal{D}_n^2$ are $\mathbf{S}_{12}(n)$-secure computationally indistinguishable and $\mathcal{D}_n^2$ and $\mathcal{D}_n^3$ are $\mathbf{S}_{23}(n)$-secure computationally indistinguishable then $\mathcal{D}_n^1$ and $\mathcal{D}_n^3$ are $\mathbf{S}_{13}(n)$-secure computationally indistinguishable, where

$$\mathbf{S}_{13}(n) = \Omega(\min\{\mathbf{S}_{12}(n), \mathbf{S}_{23}(n)\}/n^{\mathcal{O}(1)}).$$

Exercise 28 72
Describe an oracle adversary S such that if A is an adversary for $\mathcal{D}_n$ and $\mathcal{E}_n$ with time-success ratio $\mathbf{R}'(nk(n))$ then S^A is an adversary for $\mathcal{D}_n'$ and $\mathcal{E}_n'$ with time-success ratio $\mathbf{R}(n)$, where $\mathbf{R}(n) = n^{\mathcal{O}(1)} \cdot \mathcal{O}(\mathbf{R}'(nk(n)))$.

Exercise 29 72
Let $\mathcal{D}_n : \{0,1\}^n \to \{0,1\}^{\ell(n)}$ and $\mathcal{E}_n : \{0,1\}^n \to \{0,1\}^{\ell(n)}$ be **P**-samplable probability ensembles with common security parameter n. Let $f : \{0,1\}^{\ell(n)} \to \{0,1\}^{p(n)}$ be a **P**-time function ensemble. Let

$X \in_{\mathcal{D}_n} \{0,1\}^{\ell(n)}$ and $Y \in_{\mathcal{E}_n} \{0,1\}^{\ell(n)}$. Let $f(X)$ and $f(Y)$ be **P**-samplable probability ensembles with common security parameter n. Describe an oracle adversary S such that if A is an adversary for $f(X)$ and $f(Y)$ with time-success ratio $\mathbf{R}'(n)$ then S^A is an adversary for $\mathcal{D}_n$ and $\mathcal{E}_n$ with time-success ratio $\mathbf{R}(n)$, where $\mathbf{R}(n) = n^{\mathcal{O}(1)} \cdot \mathcal{O}(\mathbf{R}'(n))$.

Exercise 30 75
Prove Theorem 7.1.

Lecture 8

Exercise 31 79
Prove that for all $z > 0$, $\ln(z) \le z - 1$.

Exercise 32 (easy) 80
Let X and Y be independent random variables and let $Z = \langle X, Y \rangle$. Show that $\text{ent}(Z) = \text{ent}(X) + \text{ent}(Y)$.

Exercise 33 80
Prove the Kullback-Liebler information divergence inequality.

Exercise 34 80
Let X and Y be random variables that are not necessarily independent and let $Z = \langle X, Y \rangle$. Show that $\text{ent}(Z) \le \text{ent}(X) + \text{ent}(Y)$.

Exercise 35 81
Prove the Kraft inequality.

Exercise 36 81
Prove that for all prefix free encodings f and for all random variables X,

$$\mathbf{E}_X[\|f(X)\|] \ge \text{ent}(X).$$

Exercise 37 84
Prove that for any random variable X,

$$\text{ent}_{\text{Ren}}(X)/2 \le \text{ent}_{\min}(X) \le \text{ent}_{\text{Ren}}(X) \le \text{ent}(X).$$

Lecture 9

Exercise 38 (easy) 92
Let $X \in_{\mathcal{U}} \{0,1\}^n$. Show that

$$\text{infor}_X(x) - \text{infor}_{f(X)}(f(x)) = \log(\sharp \text{pre}_f(f(x))).$$

Let $f(x)$ be a $\sigma(n)$-regular function ensemble. Let $X \in_{\mathcal{U}} \{0,1\}^n$, $Z \in_{\mathcal{U}} \{0,1\}^n$ and $B \in_{\mathcal{U}} \{0,1\}$. Let $\mathcal{D}_n = \langle f(X), X \odot Z, Z \rangle$ and let $\mathcal{E}_n = \langle f(X), B, Z \rangle$. Show that $\mathcal{D}_n$ and $\mathcal{E}_n$ are at most $(1/\sqrt{\sigma(n)})$-statistically distinguishable.

Lecture 10

Suppose $f : \{0,1\}^n \to \{0,1\}^{\ell(n)}$ is a one-way function and $\mathrm{rank}_f(x)$ is a **P**-time function ensemble. Prove that $g(x) = \langle f(x), \mathrm{rank}_f(x) \rangle$ is a one-way one-to-one function.

Let $f : \{0,1\}^n \to \{0,1\}^{\ell(n)}$ be a one-way function and suppose $d(f(x))$ is a **P**-time function ensemble. Use Theorem 10.1 to show that there is a poly-preserving reduction from f to a pseudorandom generator.

Show that there is a linear-preserving reduction from a pseudorandom generator to a one-way function.

Design a linear-preserving (or poly-preserving) reduction from an arbitrary one-way function to a pseudorandom generator.

Lecture 11

Given that $p = \Pr[B_1 = 1]$, prove that the maximum possible covariance $p(1 - p)$ is achieved when $B_1 = B_0$.

Prove that if g is a pseudorandom generator then the stream private key cryptosystem described above is secure against passive attack. The reduction should be linear-preserving.

Prove that the previously described stream private key cryptosystem is secure against simple chosen plaintext attack. The reduction should be linear-preserving.

Prove that the stream private key cryptosystem described previously based on a pseudorandom generator is secure against chosen plaintext attack. The reduction should be linear-preserving.

Lecture 12

Watch the movie "Midway".

Prove that the block private key cryptosystem based on pseudorandom function generator f described above is secure against chosen plaintext attack. The reduction from the pseudorandom function generator f to the secure block private key cryptosystem should be linear-preserving.

Show there is a linear-preserving reduction from a pseudorandom function generator to a pseudorandom generator.

Let $g(x)$ be a pseudorandom generator that doubles the length of its input. Let $X \in_{\mathcal{U}} \{0,1\}^n$. Describe how to use g to construct two sequences of random variables

$$Y_0(X), \ldots, Y_{2^n-1}(X) \in \{0,1\}^n$$

and

$$Z_0(X), \ldots, Z_{2^n-1}(X) \in \{0,1\}^n$$

with the property that it is easy to compute Y forward from Z but hard to compute Y backward from Z.

Lecture 13

Let $F_0 \in_{\mathcal{U}} \mathbf{Perm}{:}\{0,1\}^n \to \{0,1\}^n$ and $F_1 \in_{\mathcal{U}} \mathbf{Fnc}{:}\{0,1\}^n \to \{0,1\}^n$. Show that any adversary A that makes at most m queries to an oracle has success probability at most $m^2/2^n$ for distinguishing F_0 and F_1.

Lecture 14

Prove or disprove that $\langle g, \bar{g} \rangle$ is a pseudorandom invertible permutation generator.

Suppose that P is allowed to produce a pair of message blocks that are equal to a message block produced by M in the definition of chosen plaintext attack given above. Describe a P, M and A that have a constant success probability, independent of the security of the encryption function used.

Lecture 15

Lecture 16

where the adversary A first sees $y \in_\mathcal{U} \{0,1\}^n$ and then tries to produce a pair $x, x' \in \{0,1\}^{d(n)}$, such that $x \neq x'$ but $g_y(x) = g_y(x')$?

Exercise 62 (easy) 158
Show that if f is a one-way permutation then $g^{t(n)}$ is a universal one-way hash function. The reduction should be linear-preserving.

Exercise 63 160
Design a universal one-way hash function $h_y(x)$ that maps $x \in \{0,1\}^{m(n)}$ down to $h_y(x) \in \{0,1\}^n$ such that the length of y is $\mathcal{O}(n^2 \cdot \log(m(n)/n))$.

Lecture 17

Exercise 64 165
Show that the Square root signature scheme is not worst case secure.

Exercise 65 165
ow that if factoring is hard then the Square root signature scheme is distributionally secure with respect to the message distribution that is uniform on Q_v. The reduction should be linear-preserving.

Exercise 66 166
Formally prove Theorem 17.1.

Exercise 67 172
Prove that if f is a one-way function then the One bit signature scheme is worst case secure. The reduction should be linear-preserving.

Exercise 68 172
Show how to construct a multiple messages signature scheme that uses a balanced tree structure, i.e., the length of a path in the data structure to sign the ith is of length at most $\log(i)$.

Exercise 69 173
Show how to build a multiple messages signature scheme such that the memory needed by the signer is $n^{\mathcal{O}(1)}$, and doesn't depend on the number of messages signed.

Lecture 18

Exercise 70 (easy) 175
Prove that **NP** = restricted **IP**.

Exercise 71 175
Prove that you can increase the probability spread between the completeness and soundness conditions by repeating the protocol several

times.

Show that GNI has a perfect **ZKIP**.

Prove that if g is a pseudorandom generator then the hidden bit commitment protocol hides b. The reduction should be linear-preserving. Prove that the protocol is 2^n-secure committed, independent of whether or not g is a pseudorandom generator.

List of Primary Results

Lecture 8

Lecture 9

Lecture 10

Lecture 11

Lecture 12

Lecture 13

Lecture 14

Lecture 15

Lecture 16

Lecture 17

Lecture 18

Credits and History

Although it is incomplete (like many of Michelangelo's best sculptures), [34, Goldreich] is an invaluable source for understanding the conceptual and philosophical issues of cryptography, and provides a good picture of the evolution of the field as well: it is very highly recommended reading and thinking material. An excellent survey and reasonable historical overview of cryptography in general can be found in [95, Rivest]. [95, Rivest] also contains an extensive list of references, and [25, Diffie, Hellman] is a good general reference for much of the earlier work. (I am employing double pointers here to avoid repetition of effort.) [90, Papadimitriou] is a new general reference for computational complexity, and [83, Motwani, Raghavan] is a new general reference for randomized algorithms.

Preliminaries

Standard references for complexity classes are [32, Garey and Johnson] and [52, Hopcroft and Ullman].

[22, Cobham] and [27, Edmonds] were the first to focus attention on the fundamental distinction between fast and slow algorithms. [27, Edmonds] was the first to advocate that the class of problems which are in **P** (page 8) can be efficiently solved, and [28, Edmonds] introduced an informal notion analogous to **NP** (page 8). [26, Cook] made a number of outstanding contributions, including providing the definition of **NP** (page 8) as a formal complexity class, introducing the idea of polynomial time reductions for algorithms that recognize **NP** languages, introducing the concept of **NP**-completeness, and finally making the fundamental observation that the satisfiability problem is **NP**-complete. [61, Karp] showed that a large number of natural and important combinatorial problems are **NP**-complete. This paper made it clear that the result of [26, Cook] had broad theoretical and practical implications for efficient computation. This paper also provided a scientific explanation of why researchers were having difficulty developing efficient algorithms for a variety of important practical problems.

Isolated from the West in Russia, [68, Levin] independently produced a large part of the work reported in [26, Cook] and [61, Karp], although Levin's paper did not contain the full array of **NP**-completeness results found in [61, Karp], and because of this and because he was isolated from the West, his work had a much smaller effect. Although the journal publication date of Levin's work is 1973, he talked about his work in 1971 and the journal submission date is 1972. In fact, Levin knew the

NP-completeness of Tiling many years before but did not publish this. Fortunately for Leonid, he followed the advice of Kolmogorov and did publish his work when he did, even though he didn't manage to prove that graph isomorphism is **NP**-complete, the result he really wanted. This problem is still open to this day. Some aspects of [68, Levin] are better, i.e., his optimal algorithm eliminates some awkward theoretical possibilities, and his DNF minimization is probably harder than many early reductions. Levin's dearest goal was to show **NP**-completeness for the problem of finding a short fast program for producing a given string, and his DNF minimization result was a surrogate (like graph homomorphisms and embeddings **NP**-completeness results are surrogates for resolving the graph isomorphism problem). Kolmogorov twisted his hand to make him publish what he considered a weak result. According to Leonid, "It is one of the zillions of times when Kolmogorov proved to be smarter then me." However, Leonid admits (and I know from personal experience) that he has always been a reluctant writer. For many tales about this history, see [106, Trakhtenbrot].

There were several early papers on probabilistic algorithms including [73, Lovasz], [91, Rabin], [98, Schwartz], [105, Solovay and Strassen]. These papers were perhaps the first to show that probabilistic algorithms have important and interesting applications. [33, Gill] introduced the probabilistic complexity classes **RP** (page 8), **BPP** (page 8), **PP** (page 8), and **ZPP** = **RP** ∩ co-**RP**. (**RP** is called "**VPP**" in [33, Gill], where "V" stands for "verifiable" and "PP" stands for "probabilistic polynomial time".) [1, Adleman] shows that **RP** ⊆ **P/poly** (Exercise 3 on page 10) and the result **BPP** ⊆ **P/poly** (part of the same exercise) first appeared in print in a paper on random oracles by [9, Bennett and Gill]. (Apparently several people proved this result on their own after seeing the work of [1, Adleman], but didn't publish it. This observation was a minor part of [9, Bennett and Gill].)

Interesting and useful inequalities can be found in [45, Hardy, Littlewood and Polya].

Lecture 1

The one-time pad cryptosystem (page 14) was invented in 1917 by Gilbert Vernam (see [59, Kahn]). The information-theoretic basis for the proof of its security was developed by [102, Shannon] and later refined in [50, Hellman]. The idea of sending messages longer than the initially shared private key is first considered in the context of public key cryptography, where there is no shared private key. Based on preliminary work of [81, Merkle], [24, Diffie and Hellman] introduce an informal notion of a public key cryptosystem. [41, Goldwasser and Micali] develop the formal concept of a public key cryptosystem and what it means to be secure.

[15, Blum and Micali] introduce the fundamental concept of a pseudorandom generator that is useful for cryptographic (and other) applications, and gave it the significance it has today by providing the first provable construction of a pseudorandom generator based on the conjectured difficulty of a well-known and well-studied computational problem. In particular, both the definition of pseudorandom generator based on the next bit test (page 51) and the construction of a pseudorandom generator based on the difficulty of the discrete log problem (page 17) can be found in [15, Blum and Micali].

[107, Yao] introduces the now standard definition of a pseudorandom generator (page 15), and shows an equivalence between the this definition and the next bit test (page 51 introduced in [15, Blum and Micali]. The standard definition of a pseudorandom generator introduced by [107, Yao] is based on the fundamental concept of computational indistinguishability introduced previously in [41, Goldwasser and Micali]. [107, Yao] also shows how to construct a pseudorandom generator from any one-way permutation (see the credits for Lecture 6 below for more discussion of this result).

Another important observation of [107, Yao] is that a pseudorandom generator can be used to reduce the number of random bits needed for any probabilistic polynomial time algorithm, and this shows how to perform a deterministic simulation of any polynomial time probabilistic algorithm in subexponential time based on a pseudorandom generator. The results on deterministic simulation were subsequently generalized in [16, Boppana and Hirschfeld], Work related to this, which discusses pseudorandom generators for other complexity classes, includes [87, Nisan] and [88, Nisan and Wigderson].

The notion of randomness tests for a string evolved over time: from set-theoretic tests to enumerable [64, Kolmogorov], recursive and finally limited time tests. There were some preliminary works that helped motivate the concept of a pseudorandom generator including [99, Shamir].

The robust notion of a pseudorandom generator, due to [15, Blum and Micali], [107, Yao], should be contrasted with the classical methods of generating random looking bits as described in, e.g., [63, Knuth]. In studies of classical methods, the output of the generator is considered good if it passes a particular set of standard statistical tests. The linear congruential generator is an example of a classical method for generating random looking bits that pass a variety of standard statistical tests. However, [17, Boyar] and [65, Krawczyk] show that there is a polynomial time statistical test which the output from this generator does not pass.

A good starting reference that discusses the relationship between the
$\mathbf{P} = \mathbf{NP}$ question and distributions (mentioned in the note to Exercise 6
on page 16) is [71, Levin].

Many papers have used specific conjectured one-way functions directly
as the basis for a cryptosystem. Examples include [15, Blum and Mi-
cali], [96, Rivest, Shamir and Adleman], [12, Blum, Blum and Shub], [3,
Alexi, Chor, Goldreich and Schnorr], [55, Impagliazzo and Naor]. [15,
Blum and Micali] use the discrete log problem (page 17) as the basis
for the pseudorandom generator they describe. The security of the RSA
cryptosystem ([96, Rivest, Shamir and Adleman]) is based on a version
of the root extraction problem (page 17). [41, Goldwasser and Micali]
introduce a cryptosystem based on the quadratic residuosity problem
(page 150). [107, Yao] and [12, Blum, Blum and Shub] also base a
cryptosystem on the quadratic residuosity problem. [3, Alexi, Chor,
Goldreich and Schnorr] show the security of the same cryptosystem can
be based on the weaker assumption that the factoring problem (page 17)
is hard. [55, Impagliazzo and Naor] construct a pseudorandom generator based on the difficulty of the subset sum problem (page 18). [38,
Goldreich, Krawczyk and Luby] and [60, Kaliski] show how to construct
a pseudorandom generator based on the difficulty of specific problems
not described in the monograph.

Lecture 2

The definitions given in this lecture are fairly standard in the literature,
except for the following.

Traditionally, only one type of physical security is considered for a party,
i.e., the memory, random number generator and computational unit are
all considered completely inaccessible to any adversary. The two levels
of physical security adopted in this monograph in large part is the result
of writing the monograph. A preliminary model of two levels of physical
security is described in [51, Herzberg and Luby], although the model
finally adopted in this monograph is somewhat different and arguably
more natural than the definition given there. All definitions given with
public input appear first in this monograph.

The general definition of a one-way permutation (page 28) is from [107,
Yao], and the generalization given here to a one-way function (page 27)
is due to [69, Levin]. The concept of a one-way function is closely related
to the work described in [70, Levin].

The quantification of what it means for an adversary to break a primitive,
(as opposed to just saying a breaking adversary runs in polynomial time
and has inverse polynomial success probability), the idea of using a single

achievement parameter to measure security, and the idea of focusing on the amount of security a reduction preserves have all been strongly advocated by Leonid Levin. The division of preserving properties of reductions into the three levels is introduced in this monograph.

Lecture 3

The statement of Theorem 3.1 (page 36) is due to [107, Yao], and a proof appears in [34, Goldreich]. The original strongly preserving reduction from a weak one-way permutation to a one-way permutation (the reduction described in the lecture which starts on page 46) is from [36, Goldreich, Impagliazzo, Levin, Venkatesan and Zuckerman]. Their work in turn is based on expander graph construction results due [78, Margulus] and extended in [30, Gaber and Galil]. The part of the analysis that uses the fact that a random walk on an expander graph converges quickly to the uniform distribution is from [2, Ajtai, Komlós and Szemerédi]. For more details on this, see either [23, Cohen and Wigderson] or [56, Impagliazzo and Zuckerman].

The intermediate theorems (Theorem 3.2 on page 36 and Theorem 3.3 on page 43) are partly a result of writing this monograph, and are first announced in [51, Herzberg and Luby]. These simpler strongly preserving reductions are possible in large part because of the distinction made between the private and public input as described in Lecture 2. The unifying framework for all the proofs, and in particular the Forward to Reverse Theorem (page 38), first appear in this monograph.

Lecture 4

The Self-Reducible Theorem (page 49) is due to [15, Blum and Micali]. [15, Blum and Micali] introduced the concept of a pseudorandom generator based on next bit unpredictability (page 51). [107, Yao] introduced the definition of a pseudorandom generator used in this monograph (page 50) and proved Theorem 4.1 (page 51). As described in [16, Boppana and Hirschfeld], the Stretching Theorem (page 52) is based on the work of [39, Goldreich, Goldwasser and Micali]. Exercise 12 (page 54) and its solution for the **RP** case is from [107, Yao], and the extension to the **BPP** case is due to [16, Boppana and Hirschfeld].

Lecture 5

The explicit identification of the general paradigm was developed in general terms in a series of papers [20, Chor and Goldreich], [74, Luby], [4, Alon,Babai and Itai]. The first to give an example of random variables which are pairwise independent but not mutually independent is Bernstein in 1945 (see [29, Feller], page 126). His example consists of three

$\{0, 1\}$-valued random variables. [67, Lancaster] generalizes this example to $n - 1$ pairwise independent variables on n sample points, giving constructions involving the Hadamard matrix and Latin squares. [57, Joffe] introduces the Modulo Prime Space (page 57) and [58, Joffe] generalizes this to k-wise independent variables (see Exercise 13 on page 57). [89, O'Brien] discusses generating pairwise independent random variables with additional properties.

[19, Carter and Wegman] introduce the idea of a universal hash function, which turns out to be equivalent to generating small independence random variables (see the credits for Lecture 8 below). [104, Sipser] was the first to see that these hash functions were useful for obtaining results in computational complexity. The Inner Product Space is defined and used in [62, Karp and Wigderson], and this is based on Incomplete Block Designs (see [46, Hall]). Method 2 of witness sampling (page 59 is from [20, Chor and Goldreich]. The application to the vertex partition problem (page 61) is from [75, Luby]. For other applications of this paradigm, see [4, Alon, Babai and Itai], [74, Luby], [75, Luby], [10, [Berger and Rompel], [82, Motwani and Naor].

Lecture 6

The definition of a computationally hidden but statistically meaningful bit and the realization of its importance as a basic building block for cryptographic constructions is from [15, Blum and Micali].

The construction of a hidden bit using the inner product bit (page 64, the Hidden Bit Theorem (page 65) and the Hidden Bit Technical Theorem (page 65) are all from [37, Goldreich and Levin]. The simpler proof given here of Hidden Bit Technical Theorem is due to C. Rackoff, R. Venkatesan and L. Levin, inspired by [3, Alexi, Chor, Goldreich and Schnorr]. A stronger version of Hidden Bit Theorem and Hidden Bit Technical Theorem that is linear-preserving can be found in [72, Levin]. Based on the proof ideas in [37, Goldreich and Levin], the paper [55, Impagliazzo and Naor] shows that the hardest instance of the subset sum problem is when the number of numbers is equal to the length of each number.

Using a more complicated construction, [107, Yao] was the first to show how to construct a pseudorandom generator from any one-way permutation (Exercise 19 (page 66). Some of the arguments needed in the proof were missing in [107, Yao] and were later completed by [69, Levin]. Also, [69, Levin] conjectured that a much simpler construction would work, and this eventually led to the proof that the inner product bit result mentioned above by [37, Goldreich and Levin]. Thus, the simple construction of a pseudorandom generator given in Exercise 19 (page 66) was one of the motivating forces behind the work of [37, Goldreich and Levin].

Lecture 7

The concept of computational indistinguishability (page 71) is from [41, Goldwasser and Micali].

The Theorem 7.1 (page 74) is a combination of the Stretching Theorem (page 52) due to [39, Goldreich, Goldwasser and Micali] and the Hidden Bit Theorem (page 65) due to [37, Goldreich and Levin].

Hidden Bit Corollary (page 73), Many Hidden Bits Theorem (page 75) and Many Hidden Bits Technical Theorem (page 75) are all from [37, Goldreich and Levin].

[49, Håstad, Schrift and Shamir] shows that the discrete log problem as a one-way function hides a constant fraction of the input bits simultaneously, without using the inner product bit. Thus, for this problem, a very strong version of the Many Hidden Bits Theorem is true. This can be used to immediately get a pseudorandom generator that stretches by a constant factor based on one application of the discrete log problem, and the reduction is linear-preserving. To be able to do the same thing for any one-way function would be an important breakthrough. This is related to the research problem 4 on page 104.

Lecture 8

The concept of entropy (page 79) is due to C. E. Shannon. The original paper [101, Shannon] is most easily found as [103, Shannon]. A good reference on entropy and the inequalities stated in this lecture is [31, Gallager], and a more accessible introduction to this material can be found in [5, Ash]. The importance of Renyi entropy in the context of cryptographic reductions was noted by [21, Chor and Goldreich]. The fundamental concept of computational entropy (page 82) is from [107, Yao]. The definition of a pseudoentropy generator (page 83) and Theorem 8.1 (page 83) is from [53, Impagliazzo, Levin and Luby]. Universal hash functions (page 84) were introduced in [19, Carter and Wegman]. They use universal hash functions in place of totally independent hash functions. The two properties universal hash functions they use is their succinct description and their pairwise independent randomness properties, and thus this work fits into the paradigm discussed in Lecture 5. The Smoothing Entropy Theorem is from [53, Impagliazzo, Levin and Luby]. Previously, [79, McInnes] proved a lemma related to the Smoothing Entropy Theorem, and independently, [8, Bennett, Brassard and Robert] proved a similar lemma. Related versions of this lemma were found earlier, e.g., see [104, Sipser]. [56, Impagliazzo and Zuckerman] describe some other applications of this lemma.

Lecture 9

Theorem 9.1 (page 88) is due to [38, Goldreich, Krawczyk and Luby]. The proof given here of Theorem 9.1 and Theorem 9.2 (page 90) is based on [37, Goldreich and Levin] and [53, Impagliazzo, Levin and Luby]. [38, Goldreich, Krawczyk and Luby] were the first to identify regular function ensembles as being important in the context of one-way functions: they provide natural examples of conjectured one-way regular functions and prove a version of Theorem 9.3 (page 93). [38, Goldreich, Krawczyk and Luby] was also the first paper to introduce the idea of using hashing to smooth the image distribution of a one-way function (similar to the Smoothing Entropy Theorem) in the context of constructing a pseudorandom generator from a one-way function. The proof of Theorem 9.3 given here is again based on [37, Goldreich and Levin] and [53, Impagliazzo, Levin and Luby].

Lecture 10

All of the material in this lecture is from [53, Impagliazzo, Levin and Luby], including Theorem 10.1 (page 96), the definition of a false entropy generator (page 98), Theorem 10.2 (page 99), and Theorem 10.3 (page 101). The Shannon to Renyi Theorem is similar to a theorem sometimes called the Asymptotic Equipartition Theorem, which was proved in [101, Shannon] for i.i.d. random variables and proved more generally for stationary ergodic processes in [18, Breiman] and [80, McMillan]. The overall non-uniform reduction from any one-way function to a pseudorandom generator is due to [53, Impagliazzo, Levin and Luby]. A uniform reduction is given in [47, Håstad]. [53, Impagliazzo, Levin and Luby] and [47, Håstad] are combined for the journal version.

It is interesting to note the work that has been done to show implications in the other direction, i.e., that the existence of other cryptographic primitives imply the existence of a one-way function. Simple examples of these "reverse direction" reductions are Exercise 42 (page 104) and Exercise 49(page 126). [54, Impagliazzo and Luby] show that existence of a number of other primitives, including pseudorandom generators, bit commitment and private key cryptography, imply the existence of one-way functions. [97, Rompel] shows that a secure digital signature scheme implies the existence of a one-way function. A related result can be found in [35, Goldreich], which shows that the existence of pseudorandom generators is equivalent to the existence of a pair of **P**-samplable distributions which are computationally indistinguishable but statistically very different.

Lecture 11

All of the definitions relating to stream cryptosystems in this lecture are derivatives of definitions that have been around for years, although the particular definitions were invented solely for purposes of this monograph.

Lecture 12

All of the definitions relating to block cryptosystems in this lecture are derivatives of definitions that have been around for years, although the particular definitions were invented solely for purposes of this monograph.

The definition of a pseudorandom function generator (page 122), Theorem 12.1 (page 123), the construction of a block cryptosystem from a pseudorandom function generator (page 126) and the solution to Exercise 48 (page 126) are all due to [39, Goldreich, Goldwasser and Micali].

Exercise 50 (page 126) and its solution is from [11, Blum].

The story about the movie "Midway" (page 120) is a loose transcription of a version I remember hearing from Charlie Rackoff. In the meantime, my uncle tells me that in fact Midway Island was not liberated till towards the end of the war. I did not manage to solve Exercise 47 (page 121).

Lecture 13

DES (page 128) was designed solely by IBM, and the acceptance as a standard was handled by the National Bureau of Standards [86, NBS]. The remaining part of the lecture, including the formal concept of a pseudorandom invertible permutation generator (page 129) and the Permutation Technical Theorem (page 131), is from [76, Luby and Rackoff].

Lecture 14

This lecture, including Permutation Theorem (page 138), Exercise 54 (page 142) and its solution, the definition of super pseudorandom invertible permutation generator (page 143), Exercise 55 (page 143) and its solution, and Exercise 56 (page 143) and its solution, are all from [76, Luby and Rackoff].

Lecture 15

[24, Diffie and Hellman] introduce the concept of a trapdoor one-way function (page 146). This paper started a revolution in the area of public key cryptography. [96, Rivest, Shamir and Adleman] give the first practical example of a conjectured trapdoor one-way function based on the root extraction problem (page 147). Theorem 15.1 (page 147) is due to [94, Rabin]. [41, Goldwasser and Micali] introduce the concept of a trapdoor one-way predicate (page 149), and the definition of a one-way predicate (page 149) is derived from that. [41, Goldwasser and Micali] also introduces the quadratic residuosity problem (page 150) as an example of a conjectured trapdoor one-way predicate and construct a probabilistic public key cryptosystem based on this, which is the public key bit cryptosystem described on page 151, and prove it is secure (Exercise 59 on page 151). The construction of a public key block cryptosystem on page 151 and the proof of its security (Exercise 60 on page 152) was pointed out to Moti Yung by Shimon Even. The private factorization exchange protocol (page 152) and its analysis is due to [41, Goldwasser and Micali]. The version of the exchange protocol that works in one round when $p = q = 3 \mod 4$ (page 153) is from [13, Blum].

Lecture 16

All material from this lecture, including the definition of a universal one-way hash function (page 154), Theorem 16.1 (page 157), Theorem 16.2 (page 159), Exercise 63 (page 160) and its solution are from [85, Naor and Yung]. [97, Rompel] gives a construction of a universal one-way hash function based on any one-way function.

Lecture 17

The idea of a one-message digital signature that is distributionally secure (page 163), the square root signature scheme (page 165) and Exercise 65 and its solution are all from [92, Rabin] and [93, Rabin]. The one bit signature scheme (page 165) is from [85, Naor and Yung]. [66, Lamport] introduces the idea of linking together the blocks (called a "tagging" scheme) that is used in the many message signature scheme (page 167). This idea is also attributed to W. Diffie. The first digital signature scheme proved secure against adaptive chosen message attack (based on claw-free, factoring special case) is from [42, Goldwasser, Micali and Rivest]. [7, Bellare and Micali] gives a construction based on a trapdoor one-way permutation. The construction given here and Theorem 17.1 (page 171) are due to [85, Naor and Yung]. A construction based on any one-way function can be obtained using the result of [97, Rompel] discussed in the credits for Lecture 16. Exercises 68 (page 172) and 69 (page 173) are from [85, Naor and Yung].

Lecture 18

[43, Goldwasser, Micali and Rackoff] introduce the complexity class **IP** (page 175). [43, Goldwasser, Micali and Rackoff] also introduces the idea of quantifying how much knowledge is contained in a proof and formalize this by introducing the concept of a **ZKIP** (page 177). They also give the first examples of such proofs for languages such as quadratic residuosity (page 150). [6, Babai] introduced the complexity class AM (AM stands for Arthur-Merlin) independently of [43, Goldwasser, Micali and Rackoff] and [41, Goldwasser and Sipser] prove that **IP=AM**. However, the definitions used for AM are not directly applicable for cryptographic purposes.

The series of results showing the outstanding result **IP = PSPACE** are from [77, Lund, Fortnow, Karloff and Nisan] and from [100, Shamir].

Both Theorem 18.1 (page 176) and Theorem 18.2 (page 178), from [40, Goldreich, Micali and Wigderson], concern the celebrated graph isomorphism problem (see the credits for the Preliminaries above). It is not known if GNI $\in$ **NP**, whereas it is clear that **NP** $\in$ **IP**. Theorem 18.1 shows that GNI $\in$ **IP**, and this is the first non-trivial complexity result shown about GNI. It is clear that GI $\in$ **NP**, but whether GI $\in$ **BPP** is still open. One way to think about perfect **ZKIP** is that it is an extension of **BPP**, and Theorem 18.2 shows the non-trivial result that GI has a perfect **ZKIP**.

The fundamental idea of bit commitment (page 181, a bit commitment protocol and its use to toss coins over the phone (page 181) are all introduced in [14, Blum]. The particular bit commitment scheme given on page 182 is from [84, Naor], as well a the proof that it is secure (Exercise 73 on page 183).

[40, Goldreich, Micali and Wigderson] the original proof of the foundational result that every language in **NP** has a **ZKIP**, based on the **NP**-complete problem three-colorability of graphs. The proof given here based on the Hamilton cycle problem (Theorem 18.3 on page 183) is due to M. Blum.

References

Abbreviations :

- **STOC**: Proceedings of the ACM Symposium on Theory of Computing

- **FOCS**: Proceedings of the IEEE Foundations of Computer Science

References

[1] L. Adleman, "Two Theorems on Random Polynomial Time", **FOCS**, 1978, pp. 75–83.

[2] M. Ajtai, A. Komlós and E. Szemerédi, "Deterministic Simulation in LOGSPACE", **STOC**, 1987, pp. 132–140.

[3] W. Alexi, B. Chor, O. Goldreich and C. Schnorr, "RSA/Rabin Functions: Certain Parts are as Hard as the Whole", *SIAM J. on Computing*, Vol. 17, No. 2, April 1988, pp. 194–209.

[4] N. Alon, L. Babai, A. Itai, "A Fast and Simple Randomized Parallel Algorithm for the Maximal Independent Set Problem", *Journal of Algorithms*, Vol. 7, 1986, pp. 567–583.

[5] R. B. Ash, *Information Theory*, Dover Publishers, 1965.

[6] L. Babai, "Trading Group Theory for Randomness", **STOC**, 1985, pp. 421–429.

[7] M. Bellare and S. Micali, "How to sign given any trapdoor permutation", *J. of the ACM*, Vol. 39, No. 1, January 1992, pp. 214-233. A preliminary version appears in **STOC**, 1988, pp. 32–42.

[8] C. Bennett, G. Brassard and J. Robert, "Privacy Amplification by Public Discussion", *Siam J. on Computing*, Vol. 17, No. 2, 1988, pp. 210–229.

[9] C. Bennett and J. Gill, "Relative to a random oracle A, $\mathbf{P}^A \neq \mathbf{NP}^A \neq co - \mathbf{NP}^A$ with probability one", *Siam J. on Computing*, Vol. 10, 1981, pp. 96–113.

[10] B. Berger, and J. Rompel, "Simulating $(\log^c n)$-wise Independence in *NC*", *J. of the ACM*, Vol. 38, No. 4, Oct. 1991, pp. 1026–1046. A preliminary version appears in **FOCS**, 1989, pp. 1–7.

[11] A. Blum, "Separating distribution-free and mistake-bound learning models over the Boolean domain", *SIAM J. on Computing.* Vol 23, No. 5, 1994, pp. 990–1000. A preliminary version appears in **FOCS**, 1990, pp. 211–218.

[12] L. Blum, M. Blum and M. Shub, "A simple unpredictable pseudo-random generator", *SIAM J. on Computing*, Vol. 15, No. 2, 1986, pp. 364–383.

[13] M. Blum, "How to exchange (secret) keys", *ACM Trans. Comput. Systems*, Vol. 1, 1983, pp. 175–193.

[14] M. Blum, "Coin flipping by telephone: a protocol for solving impossible problems", *Proceedings 24th IEEE Spring Computer Conf.*, COMPCON, 1982, pp. 133–137.

[15] M. Blum and S. Micali, "How to Generate Cryptographically Strong Sequences of Pseudo-Random Bits", *SIAM J. on Computing*, Vol. 13, 1984, pp. 850–864. A preliminary version appears in **FOCS**, 1982, pp. 112–117.

[16] R. Boppana and R. Hirschfeld, "Pseudo-random generators and complexity classes", **Advances in Computer Research**, Vol. 5, 1989, editor S. Micali, JAI Press, pp. 1–26.,

[17] J. Boyar, "Inferring Sequences Produced by Pseudo-Random Number Generators", *J. of the ACM*, Vol. 36, No. 1, 1989, pp.129–141.

[18] L. Breiman, "The individual ergodic theorems of information theory", *Ann. Math. Stat.*, Vol. 28, 1957, pp. 809–811.

[19] J. Carter and M. Wegman, "Universal Classes of Hash Functions", *JCSS*, Vol. 18, 1979, pp. 143–154.

[20] B. Chor and O. Goldreich, "On the Power of Two Point Based Sampling", *Journal of Complexity*, Vol. 5, 1989, pp. 96–106.

[21] B. Chor and O. Goldreich, "Unbiased Bits from Sources of Weak Randomness and Probabilistic Communication Complexity", *SIAM J. on Computing*, Vol. 17, No. 2, April 1988, pp. 230–261. A preliminary version appears in **FOCS**, 1985, pp. 429–442,

[22] A. Cobham, "The intrinsic computational difficulty of functions," *Proceedings International Congress for Logic Methodology and Philosophy of Science*, 1964, North Holland, Amsterdam, pp. 24–30.

[23] A. Cohen and A. Wigderson, "Dispersers, Deterministic Amplification, and Weak Random Sources", **FOCS**, 1989, pp. 14–19.

[24] W. Diffie and M. Hellman, "New directions in cryptography", *IEEE Trans. Inform. Theory*, Vol. 22, 1976, pp. 644–654.

[25] W. Diffie and M. Hellman, "Privacy and Authentication: An Introduction to Cryptography", *Proceedings of the IEEE*, Vol. 67, March 1979, pp. 397–427.

[26] S.A. Cook, "The complexity of theorem-proving procedures", **STOC**, 1971, pp. 151–158.

[27] J. Edmonds, "Paths, trees and flowers", *Canad. J. Math.*, Vol. 17, 1965, pp. 449-467.

[28] J. Edmonds, "Minimum partition of a matroid into independent sets", *J. Res. Nat. Bur. Standards Sect. B*, Vol. 69, pp. 67–72.

[29] W. Feller, **An Introduction to Probability Theory and Its Applications**, Vol. 1, Third edition, 1968, John Wiley and Sons, publishers.

[30] O. Gaber and Z. Galil, "Explicit Constructions of Linear-Sized Superconcentrators", *JCSS*, Vol. 22, 1981, pp. 407–420.

[31] R. G. Gallager, **Information Theory and Reliable Communication**, Wiley Publishers, 1968.

[32] M. Garey and D. Johnson, **Computers and Intractability: A Guide to the Theory of NP-Completeness**, W. H. Freeman and Company Publishers, 1979.

[33] J. Gill, "Computational Complexity of Probabilistic TMs", *SIAM J. on Computing*, Vol. 6, 1977, pp. 675–695.

[34] O. Goldreich, "Foundations of Cryptography", Class notes from a course taught in the spring of 1989 at the Technion, Haifa, Israel. There is a slightly updated version that appears as a monograph published by the Weizmann Institute of Science and dated February 23, 1995 with the slightly modified title "Foundations of Cryptography (Fragments of a Book)".

[35] O. Goldreich, "A Note on Computational Indistinguishability", *ICSI Tech Report TR-89-051*, July 1989.

[36] O. Goldreich, R. Impagliazzo, L. Levin, R. Venkatesan and D. Zuckerman, "Security Preserving Amplification of Hardness", **FOCS** 1990, pp. 318–326.

[37] O. Goldreich and L. Levin, "A Hard-Core Predicate for any One-way Function", **STOC**, 1989, pp. 25–32.

[38] O. Goldreich, H. Krawczyk and M. Luby, "On the Existence of Pseudorandom Generators", *SIAM J. on Computing*, Vol. 22, No. 6, December, 1993, pp. 1163–1175. A preliminary version appears in **FOCS**, 1988, pp. 12–24,

[39] O. Goldreich, S. Goldwasser and S. Micali, "How to Construct Random Functions", *J. of the ACM*, Vol. 33, No. 4, 1986, pp. 792–807. A preliminary version appears in **FOCS**, 1984, pp. 464–479.

[40] O. Goldreich, S. Micali, and A. Wigderson, "Proofs that Yield Nothing But their Validity or All Languages in NP have Zero-Knowledge Proofs", *J. of the ACM*, Vol. 38, No. 3, July 1991, pp. 691–729. A preliminary version appears in **FOCS**, 1986, pp. 174–187.

[41] S. Goldwasser and S. Micali, "Probabilistic Encryption", *J. of Computer and System Sci.*, Vol. 28, 1984, pp. 270–299. A preliminary version appears in **STOC**, 1982, pp. 365–377.

[42] S. Goldwasser, S. Micali, R. Rivest, "A secure digital signature scheme", *SIAM J. on Computing*, Vol. 17, No. 2, 1988, pp. 281–308.

[43] S. Goldwasser, S. Micali and C. Rackoff, "The Knowledge Complexity of Interactive Proof Systems," *SIAM J. on Computing*, Vol. 18, No. 1, 1989, pp. 186–208. A preliminary version appears in **STOC**, 1985, pp. 291–304.

[44] S. Goldwasser, M. Sipser, "Private coins versus public coins in interactive proof systems", **STOC**, 1986, pp. 59–68.

[45] Hardy, Littlewood and Polya, **Inequalities**, Second Edition, Cambridge University Press, 1989 printing.

[46] M. Hall Jr., **Combinatorial Theory**, 1967, Blaisdell, Waltham, Massachusetts.

[47] J. Håstad, "Pseudo-Random Generators under Uniform Assumptions", **STOC**, 1990, pp. 395–404.

[48] J. Håstad, R. Impagliazzo, L. Levin and M. Luby, "Construction of a pseudo-random generator from any one-way function", *ICSI Technical Report*, No. 91-068, submitted to *SICOMP*.

[49] J. Håstad, A.W. Schrift and A. Shamir, "The Discrete Logarithm Modulo a Composite Hides $\mathcal{O}(n)$ Bits", *JCSS*, Vol. 47, No. 3, December 1993, pp. 376–404. A preliminary version appears in **STOC**, 1990, pp. 405–415.

[50] M. E. Hellman, "An extension of Shannon theory approach to cryptography", *IEEE Trans. Infor. Theory*, Vol. 23, 1977, pp. 289–294.

[51] A. Herzberg and M. Luby, "Public Randomness in Cryptography", proceedings of *CRYPTO 1992*, and also *ICSI technical report* TR-92-068, October, 1992.

[52] J. Hopcroft and J. Ullman, *Introduction to Automata Theory, Languages and Computation*, Addison-Wesley publishing company, 1979.

[53] R. Impagliazzo, L. Levin and M. Luby, "Pseudo-random number generation from one-way functions", **STOC**, 1989, pp. 12–24.

[54] R. Impagliazzo and M. Luby, "One-way Functions are Essential for Complexity Based Cryptography," **FOCS**, 1989. pp. 230–235.

[55] R. Impagliazzo and M. Naor, "Efficient cryptographic schemes provably as secure as subset sum", *Technical Report CS93-12*, Weizmann Institute, 1993, accepted to J. of Cryptology, 1995. A preliminary version appears in **FOCS**, 1989, pp. 236–241.

[56] R. Impagliazzo and D. Zuckerman, "How to Recycle Random Bits", **FOCS**, 1989, pp. 248–253.

[57] A. Joffe, "On a Sequence of Almost Deterministic Pairwise Independent Random Variables", *Proceedings Amer. Math. Soc.*, 1971, 29, pp. 381–382.

[58] A. Joffe, "On a Set of Almost Deterministic k-Independent Random Variables", *The Annals of Probability*, 1974, Vol. 2, No. 1, pp. 161–162.

[59] D. Kahn, *The Codebreakers*, Macmillan, New York, 1967.

[60] B.S. Kaliski, "A pseudo-random bit generator based on elliptic curves", *Advances in Cryptology, CRYPTO 86*, Lecture Notes in Computer Science, Vol. 263, Springer, Berlin, 1987, pp. 84–103.

[61] R. Karp, "Reducibility among combinatorial problems", *Complexity of Computer Computations*

[62] R. Karp and A. Wigderson, "A Fast Parallel Algorithm for the Maximal Independent Set Problem", **STOC**, 1984, pp. 266–272.

[63] D. Knuth, **Semi-Numerical Algorithm, The Art of Computer Programming**, Addison-Wesley, Second Edition, Vol. 2, 1981.

[64] A. N. Kolmogorov, "Three Approaches to the Concept of the Amount of Information", *Probl. Inf. Transm.*, Vol. 1, No. 1, 1965.

[65] H. Krawczyk, "How to Predict Congruential Generators", *J. of Algorithms*, Vol. 13, 1992. pp. 527–545.

[66] L. Lamport, "Constructing digital signatures from one-way functions", SRI intl. CSL-98, October 1979.

[67] H. Lancaster, "Pairwise Statistical Independence", *Ann. Math. Statis.*, 1965, 36, pp. 1313–1317.

[68] L. Levin, "Universal sorting problems", *Problemy peredaci Informacii*, Vol. 9, 1973, pp. 115–116 (in Russian). English translation in *Problems of Information Transmission*, Vol. 9, 1973, pp. 265–266.

[69] L. Levin, "One-way functions and pseudorandom generators", *Combinatorica*, Vol. 7, No. 4, 1987, pp. 357–363. A preliminary version appears in **STOC**, 1985, pp. 363–365.

[70] L. Levin, "Average Case Complete Problems", *SIAM J. on Computing* Vol. 15, No. 1, 1986, pp. 285-286. A preliminary version appears in **STOC**, 1984, p. 465.

[71] L. Levin, "Homogeneous Measures and Polynomial Time Invariants", **FOCS**, 1988, pp. 36–41.

[72] L. Levin, "Randomness and Non-determinism", *J. of Symb. Logic*, Vol. 58, No. 3, 1993, pp. 1102–1103.

[73] L. Lovasz, "On Determinants, Matchings and Random Algorithms", Proc. 2, FCT, 1979, pp. 565–574.

[74] M. Luby, "A Simple Parallel Algorithm for the Maximal Independent Set Problem," *SIAM J. on Computing*, Volume 15, No. 4, November 1986, pp. 1036–1053. A preliminary version appears in **STOC**, 1985, pp. 1–10,

[75] M. Luby, "Removing Randomness in Parallel Computation without a Processor Penalty", *JCSS*, Vol. 47, No. 2, 1993, pp. 250–286. A preliminary version appears in **FOCS**, 1988, pp. 162–173,

[76] M. Luby and C. Rackoff, "How to Construct Pseudorandom Permutations From Pseudorandom Functions", *SIAM J. on Computing*, Vol. 17, 1988, pp. 373–386. A preliminary version appears in **STOC**, 1986, pp. 356–363.

[77] C. Lund, L. Fortnow, H. Karloff and N. Nisan, "Algebraic Methods for Interactive Proof Systems", **FOCS**, 1990, pp. 2–10.

[78] Margulus, "Explicit construction of concentrators", *Problemy Peredaci Informacii 9*, No. 4, 1973, pp. 71–80, English translation in *Problems Inform. Transmission*, 1975.

[79] J. McInnes, "Cryptography Using Weak Sources of Randomness," *Tech. Report 194/87*, U. of Toronto, 1987.

[80] B. McMillan, "The basic theorems of information theory", *Ann. Math. Stat.*, Vol. 24, 1953, pp. 196–219.

[81] R.C. Merkle, "Secure communications over insecure channels", *Comm. of the ACM*, Vol. 21, 1978, pp. 294–299.

[82] R. Motwani, J. Naor and M. Naor, "The Probabilistic Method Yields Deterministic Parallel Algorithms", *JCSS*, Vol. 49, No. 3, December 1994, pp. 478–516. A preliminary version appears in **FOCS**, 1989, pp. 8–13.

[83] R. Motwani and P. Raghavan, **Randomized Algorithms**, Cambridge University Press, 1995.

[84] M. Naor, "Bit Commitment Using Pseudo-Randomness", *Journal of Cryptology*, Vol 4, 1991, pp. 151–158.

[85] M. Naor and M. Yung, "Universal Hash Functions and their Cryptographic Applications", **STOC**, 1989, pp. 33–43.

[86] National Bureau of Standards, "Announcing the data encryption standard", *Tech. Report FIPS*, Publication 46, 1977.

[87] N. Nisan, "Pseudorandom bits for constant depth circuits", *Combinatorica*, Vol. 1, 1991, pp. 63–70.

[88] N. Nisan and A. Wigderson. "Hardness vs. Randomness", *JCSS*, Vol. 49, No. 2, 1994, pp. 149–167.

[89] G. O'Brien, "Pairwise Independent Random Variables", *The Annals of Probability*, Vol. 8, No. 1, 1980, pp. 170–175.

[90] C. Papadimitriou, **Computational Complexity**, Addison-Wesley, 1994.

[91] M. Rabin, "Probabilistic Algorithms in Finite Fields", SIAM J. on Computing 9, 1980, pp. 273–280.

[92] M. Rabin, "Digitalized Signatures", *Foundation of Secure Computation*, R.A. DeMillo, D. Dobkin, A. Jones and R. Lipton, eds., Academic Press, 1977.

[93] M. Rabin, "Digitalized signatures as intractable as factorization", *Tech. Report MIT/LCS/TR-212*, MIT Lab. Comput. Sci., 1979.

[94] M. Rabin, "How to exchange secrets by oblivious transfer", *Tech. Report TR-81*, Harvard Univ., Aiken Comput. Lab., 1981.

[95] R. Rivest, "Cryptography", **Handbook of Theoretical Computer Science**, Volume A, J. van Leeuwen editor, 1990, pp. 719–755.

[96] R. Rivest, A. Shamir and L. Adleman, "A method for obtaining digital signatures and public-key cryptosystems", *Comm. of the ACM*, Vol. 21, 1978, pp. 120–126.

[97] J. Rompel, "One-way Functions are Necessary and Sufficient for Secure Signatures", **STOC**, 1990, pp. 387–394.

[98] J. Schwartz, "Fast Probabilistic Algorithms for Verification of Polynomial Identities", *J. of the ACM*, 27, 1980, pp. 701–717.

[99] A. Shamir, "On the generation of cryptographically strong pseudorandom sequences", *ACM Transactions on Computer Systems*, Vol. 1, No. 1, 1983, pp. 38–44. A preliminary version appears in the 8th ICALP and appears in **Lecture Notes on Computer Science**, 1981, Springer Verlag, pp. 544–550.

[100] A. Shamir, "IP=PSPACE", **FOCS**, 1990, pp. 11–15.

[101] C. E. Shannon, "A mathematical theory of communication", *Bell system Tech. J.*, Vol. 27, 1948, pp. 379–423.

[102] C. E. Shannon, "Communication theory of secrecy systems", *Bell system Tech. J.*, Vol. 28, 1949, pp. 657–715.

[103] C. E. Shannon and W. Weaver, **The Mathematical Theory of Communication**, U. Illinois Press, 1949.

[104] M. Sipser, "A Complexity Theoretic Approach to Randomness", **STOC**, 1983, pp. 330–335.

[105] R. Solovay and V. Strassen, "A Fast Monte-Carlo Test for Primality", *SIAM J. on Computing*, Vol. 6, 1977, pp.84–85, and *SIAM J. on Computing*, Vol. 7, p. 118.

[106] B.A.Trakhtenbrot. "A survey of Russian approaches to *Perebor* (brute-force search) algorithms", *Annals of the History of Computing*, Vol. 6, 1984, pp. 384–400.

[107] A. Yao, "Theory and Applications of Trapdoor Functions", **FOCS**, 1982, pp. 80–91.

Notation

$\{0,1\}^n$: The set of all bit strings of length n.

$\{0,1\}^{\leq n}$: The set of all bit strings of length at most n.

λ: The empty string.

0^n: The concatenation of n 0 bits.

1^n: The concatenation of n 1 bits.

$x \in \{0,1\}^n$: x is a string of n bits.

$x \in \{0,1\}^{m \times n}$: x is a m times n matrix of bits.

$\mathrm{diag}(x)$: If $x \in \{0,1\}^{n \times n}$ then $\mathrm{diag}(x) = \langle x_{1,1}, x_{2,2}, \ldots, x_{n,n} \rangle$.

$\{1,-1\}$: This is an alternative notation for a bit where the bit is either -1 or 1 instead of either 0 or 1. If $b \in \{0,1\}$ then $\bar{b} = (-1)^b$. If $x \in \{0,1\}^n$ then the ith bit of $\bar{x}$ is $\bar{x}_i$.

$S \setminus T$: The set of all elements in set S but not in set T.

$S \times T$: The set of all ordered pairs $\langle x, y \rangle$, where $x \in S$ and $y \in T$.

$\mathcal{Z}$: The set of all integers.

$J(x,z)$: The Jacobi symbol of $x \in \mathcal{Z}_z^*$, which is either -1 or 1.

$\mathcal{J}_z$: The elements of $\mathcal{Z}_z^*$ with Jacobi symbol 1, i.e., $\mathcal{J}_z = \{y \in \mathcal{Z}_z^* : J(y,z) = 1\}$.

$\mathcal{N}$: The set of all non-negative integers.

$\mathcal{Q}_z$: The set of squares mod z, i.e., $\mathcal{Q}_z = \{y^2 \bmod z : y \in \mathcal{Z}_z^*\}$.

$\bar{\mathcal{Q}}_z$: The set of non-squares mod z with Jacobi symbol 1, i.e., $\bar{\mathcal{Q}}_z = \mathcal{J}_z \setminus \mathcal{Q}_z$.

$\mathcal{R}$: The set of real numbers.

$\{i, \ldots, j\}$: The set of integers between i and j, inclusive.

$\mathcal{Z}_p$: The set of integers $\{0, \ldots, p-1\}$, where p is typically a prime. We can view $\mathcal{Z}_p$ as a group with respect to addition modulo p.

$\mathcal{Z}_p^*$: The set of integers $\{z \in \{1, \ldots, p-1\} : \gcd(z,p) = 1\}$. We can view $\mathcal{Z}_p^*$ as a group with respect to muliplication modulo p.

$\langle x, y \rangle$: This is either the concatenation of x and y or the ordered sequence of two strings x followed by y.

$f : \{0, 1\}^n \times \{0, 1\}^{\ell(n)} \to \{0, 1\}^{m(n)}$: The function ensemble f maps two inputs, one of length n and the other of length $\ell(n)$, to an output of length $m(n)$. We often write $f_x(y) = f(x, y)$ to indicate that we view f as a function of its second input for a fixed value of its first input.

x_S: If $x \in \{0, 1\}^n$ and $S \subset \{1, \ldots, n\}$ then x_S is the subsequence of bits of x indexed by S, e.g. $x_{\{1,\ldots,i\}}$ is the first i bits of x and $x_{\{i+1,\ldots,n\}}$ is all but the first i bits of x. If x is a sequence of bit strings, then x_S is the subsequence of strings indexed by S.

x_i: Either the ith bit of x or else x_i is the ith element in a list of elements.

$x \odot r$: The multiplication of bit vectors x and r over GF[2].

$|x|$: The absolute value of x.

$\|x\|$: The length of bit string x

$\sharp S$: The number of elements in set S.

$\lceil x \rceil$ The smallest integer greater than or equal to x.

$\mathrm{Pr}_X[X = x]$: The probability that X takes on the value x.

$\mathrm{E}_X[f(X)]$: The expected value of $f(X)$ with respect to X.

$x \in_{\mathcal{U}} S$: x is chosen randomly and uniformly from the set S and fixed.

$X \in_{\mathcal{U}} S$: X is a random variable distributed uniformly in the set S.

$X \in_{\mathcal{D}_n} S$: X is a random variable distributed according to the distribution $\mathcal{D}_n$ in the set S.

$|\mathrm{E}[\overline{X} \cdot \overline{Y}]|$: When $X \in_{\mathcal{U}} \{0, 1\}$ and Y is a $\{0, 1\}$-valued random variable, then this is the correlation of Y with X.

$|\mathrm{E}[X \cdot Y] - \mathrm{E}[X] \cdot \mathrm{E}[Y]|$: When X and Y are $\{0, 1\}$-valued random variables, then this is the covariance of Y with X.

$x \oplus y$: The bit by bit exclusive-or of bit strings x and y.

$\log(x)$: The logarithm base two of x.

$\ln(x)$: The natural logarithm of x.

$f(n) = \mathcal{O}(g(n))$: There is a positive constant c such that $f(n) \leq c \cdot g(n)$.

$f(n) = \Omega(g(n))$: There is a positive constant c such that $f(n) \geq c \cdot g(n)$.

$f(n) = g(n)^{\mathcal{O}(1)}$: There is a positive constant c such that $f(n) \leq g(n)^c$.

$f(n) = g(n)^{\Omega(1)}$: There is a constant $c > 0$ such that $f(n) \geq g(n)^c$.

$f(n) = g(n^{\mathcal{O}(1)})$: There is a constant $c > 0$ such that $f(n) \leq g(n^c)$.

$f(n) = g(n^{\Omega(1)})$: There is a constant $c > 0$ such that $f(n) \geq g(n^c)$.

GF[2] **and** GF[2^n]: GF[2] is the Galois field of two elements, and GF[2^n] is the Galois field on 2^n elements.

$\mathcal{D}_n$: $\mathcal{D}_n$ is typically the nth distribution in a probability ensemble.

covar(X, Y): The covariance between two random variables X and Y. This is defined as covar$(X, Y) = E[XY] - E[X] \cdot E[Y]$.

degen$_f(n)$: The degeneracy of f with respect to the uniform distribution on its inputs of length n. Defined as

$$\mathrm{degen}_f(n) = \mathrm{E}_{X \in_{\mathcal{U}} \{0,1\}^n}[\mathrm{infor}_X(x) - \mathrm{infor}_{f(X)}(f(x))].$$

pre$_f(y)$: This is the set of preimages of $y \in \mathrm{range}_f(n)$, i.e., $\mathrm{pre}_f^y(n) = \{x : f(x) = y\}$.

range$_f(n)$: This is the set of elements in the range of f, i.e., $\mathrm{range}_f(n) = \{f(x) : x \in \{0,1\}^n\}$.

$\sigma(n)$: We say $f(x)$ is a $\sigma(n)$-regular function if for each element $y \in \mathrm{range}_f(n)$, $\sharp\mathrm{pre}_f^{f(x)}(n) = \sigma(n)$.

rank$_f(x)$: The rank of $x \in \{0,1\}^n$ among all preimages of $f(x)$, i.e., $\sharp\{x' \in \mathrm{pre}_f^{f(x)}(n) : x' < x\}$.

Fnc:$S \to T$: The family of all functions from set S to set T.

Perm:$S \to S$: The family of all permutations from set S to itself.

$\mathcal{H}, \bar{\mathcal{H}}$: These are operators on functions defined on page 129

dist(X, Y): The statistical distance in the L_1 norm between X and Y.

s(n): The security parameter of a protocol, i.e., the amount of private memory used by the nth protocol.

R$(\mathbf{s}(n))$: The ratio of success a particular adversary achieves. This is the ratio of the run time over the success probability of the adversary for breaking the protocol that uses private memory of size $\mathbf{s}(n)$.

S(s(n)): The amount of security a protocol achieves, parameterized by the amount of private memory it uses.

w(n): The weakness parameter of a weak one-way function.

infor$_X(x)$: The information of string x with respect to random variable X. Defined as

$$\log(1/\Pr_X[X = x]) = -\log(\Pr_X[X = x]).$$

ent(X): The Shannon entropy of random variable X. Defined as

$$\mathrm{E}_X[\mathrm{infor}_X(X)] = \sum_{x \in \{0,1\}^n} \Pr_X[X = x] \cdot \mathrm{infor}_X(x).$$

ent$_{\mathrm{Ren}}(X)$: The Renyi entropy of random variable X. Defined as

$$-\log\left(\Pr_{X,Y}[X = Y]\right),$$

where Y is a random variable independent of X with the same distribution as X.

ent$_{\min}(X)$: The minimum entropy of random variable X. Defined as

$$\min\{\mathrm{infor}_X(x) : x \in \{0,1\}^n\}.$$

dist(X, Y): The statistical distance between the distribution defined by random variable X and that defined by random variable Y. If X and Y are both distributed in $\{0,1\}^n$, then this is defined as

$$\max\left\{\Pr_X[X \in S] - \Pr_Y[Y \in S] : S \subseteq \{0,1\}^n\right\}.$$

$T(n)$: Typically the time bound on adversaries.

$\delta(n)$: Typically the success probability of an adversary.

$d(f(x))$: $d(f(x)) = \lceil \log(\sharp\mathrm{pre}(f(x))) \rceil$.

S^A: If S is an oracle Turing machine and A is another Turing machine then S^A denotes the oracle Turing machine S with its oracle calls computed by A.

$M(X)$: If M is a Turing machine and X is a random variable then $M(X)$ denotes the random variable defined by the output of M when the input is distributed according to X.

TM: Turing machine.

Index

Milton Keynes UK
Ingram Content Group UK Ltd.
UKHW040747180824
447095UK00001B/66